❖ 49 ❖
Easy Electronic Projects for Transconductance & Norton Op Amps

Delton T. Horn

TAB BOOKS
Blue Ridge Summit, PA

Allen County Public Library
Ft. Wayne, Indiana

FIRST EDITION
FIRST PRINTING

Copyright © 1990 by TAB BOOKS
Printed in the United States of America

Reproduction or publication of the content in any manner, without express permission of the publisher, is prohibited. The publisher takes no responsibility for the use of any of the materials or methods described in this book, or for the products thereof.

Library of Congress Cataloging-in-Publication Data

Horn, Delton T.
    49 easy electronic projects for transconductance & Norton op amps / by Delton T. Horn.
      p. cm.
    Includes index.
    ISBN 0-8306-7455-1   ISBN 0-8306-3455-X (pbk.)
    1. Operational amplifiers—Amateurs' manuals. I. Title.
II. Title: Forty nine easy electronic projects for transconductance & Norton op amps.
TK9966.H663   1990
621.39'5—dc 20   90-37043
                                        CIP

TAB BOOKS offers software for sale. For information and a catalog, please contact TAB Software Department, Blue Ridge Summit, PA 17294-0850.

Questions regarding the content of this book should be addressed to:

    Reader Inquiry Branch
    TAB BOOKS
    Blue Ridge Summit, PA 17294-0214

Acquisitions Editor: Roland S. Phelps
Technical Editor: Daniel Early
Production: Katherine Brown

Contents

List of Projects *ix*

Introduction *xi*

❖ Part I ❖
Transconductance Operational Amplifiers

❖ 1 The Transconductance Operational Amplifier   *3*
What Is an OTA? 4
How an Op Amp Works 4
How an OTA Works 9
OTA Applications 10

❖ 2 The CA3080 and Other Typical OTA Devices   *13*
Specifications for the CA3080 13
The CA3080's Internal Structure 16
Compensation 21
The LM13600 Dual OTA 24

❖ 3 Amplifier Projects   *31*
Direct-coupled Differential Amplifier 32
AC-coupled Inverting Amplifier 36
Low-power Inverting Amplifier 40
Variable-gain Inverting Amplifier 43
AGC Amplifier 44
Simple VCA 47
Improved VCA 50
LM13600 VCA 51

❖ 4 Switching and Modulation Projects   *55*
Analog Switch 55
Fast-Inverting Switch 58
Schmitt Trigger 60

      Low-power Schmitt Trigger   63
      Variable-Hysteresis Schmitt Trigger   63
      Sample-and-Hold Circuit   67
      Inverting Voltage Comparator   71
      Noninverting Voltage Comparator   73
      CA3080 Amplitude Modulator   75
      LM13600 Amplitude Modulator   78
      AM Sidebands   80
      Ring Modulator   82
      Four-quadrant Multiplier   84
      Phase-Amplitude Modulator   87

❖ 5 Signal-Generating Projects   *91*
      Square-Wave Generator   91
      Variable Duty-Cycle Rectangular-Wave Generator   94
      VCO   97
      Random Music Maker   100

❖ 6 Miscellaneous OTA Projects   *105*
      Harmonics Remover   105
      Voltage-controlled Resistance   108
      Precision Current Source   110
      Low-pass VCF   111
      High-pass VCF   119

❖ Part II ❖
Norton/Amplifiers

❖ 7 The Norton Amplifier   *125*
      Basic Principles of the Norton Op Amp   125
      Biasing   127
      Inverting Circuits   128
      Noninverting Circuits   131

❖ 8 The LM3900   *135*
      Specifications   135
      Internal Structure   136
      Breaking Down the Circuitry   138

❖ 9 Amplifier Projects   *145*
      Inverting Amplifier   145
      Noninverting Amplifier   148

Differential Amplifier 151
Wide-bandwidth/High-gain Amplifier 155
High-pass Filter 157

❖10 Voltage and Current Projects — *163*

Voltage Regulator 163
Variable Voltage Regulator 165
Variable-Reference Voltage Source 167
Schmitt Trigger 168
Voltage Comparator 171
Under-Voltage Indicator 173
Fixed-Current Source 174
Simple Current Sink 177
Improved Current Sink 179

❖11 Miscellaneous Norton Amplifier Projects — *183*

Four-input AND Gate 183
Flip-Flop 185
Low-temperature Alarm 186
High-temperature Alarm 189
Square-Wave Generator 191
Alternative Square-Wave Generator 194
Narrow-width Pulse Generator 196
Variable Duty-Cycle Rectangular-Wave Generator 198
Function Generator 202
Dual LED Flasher 203

Index — *207*

List of Projects

Project	Description	Chapter/Page
	OTA Projects	
1	Direct-coupled Differential Amplifier	3/32
2	AC Coupled Inverting Amplifier	3/36
3	Low-power Inverting Amplifier	3/40
4	Variable-gain Inverting Amplifier	3/43
5	AGC Amplifier	3/44
6	Simple VCA	3/47
7	Improved VCA	3/50
8	LM13600 VCA	3/51
9	Analog Switch	4/55
10	Fast-inverting Switch	4/58
11	Schmitt Trigger	4/60
12	Low-power Schmitt Trigger	4/63
13	Variable-Hysteresis Schmitt Trigger	4/64
14	Sample-and-Hold Circuit	4/67
15	Inverting Voltage Comparator	4/71
16	Noninverting Voltage Comparator	4/73
17	CA3080 Amplitude Modulator	4/78
18	LM13600 Amplitude Modulator	4/78
19	Ring Modulator	4/82
20	Four-quadrant Multiplier	4/84
21	Phase-Amplitude Modulator	4/87
22	Square-Wave Generator	5/91
23	Variable Duty-cycle Rectangular-Wave Generator	5/94
24	VCO	5/97

Project	Description	Chapter/Page
25	Random Music Maker	5/100
26	Harmonics Remover	6/105
27	Voltage-controlled Resistance	6/108
28	Precision Current Source	6/110
29	Low-pass VCF	6/111
30	High-pass VCF	6/119

Norton Amplifier Projects

Project	Description	Chapter/Page
31	Inverting Amplifier	9/145
32	Noninverting Amplifier	9/148
33	Differential Amplifier	9/151
34	Wide-bandwidth/High-gain Amplifier	9/155
35	High-pass Filter	9/157
36	Voltage Regulator	10/163
37	Variable Voltage Regulator	10/165
38	Variable Reference Voltage Source	10/167
39	Schmitt Trigger	10/168
40	Voltage Comparator	10/171
41	Under-Voltage Detector	10/173
42	Fixed-Current Source	10/174
43	Simple Current Sink	10/177
44	Improved Current Sink	10/179
45	Four-input AND Gate	11/183
46	Flip-Flop	11/185
47	Low-temperature Alarm	11/186
48	High-temperature Alarm	11/189
49	Square-Wave Generator	11/191
50*	Alternative Square-Wave Generator	11/194
51*	Narrow-width Pulse-Generator	11/196
52*	Variable Duty-Cycle Rectangular-Wave Generator	11198
53*	Function Generator	11/202
54*	Dual LED Flasher	11/203

*bonus projects

Introduction

ARE THERE ANY ELECTRONICS EXPERIMENTERS TODAY WHO ARE NOT familiar with the operational amplifier, or *op amp*? The op amp is used in thousands of different circuits. It is almost certainly the most popular type of IC available and definitely one of the most versatile. The op amp is the most general purpose of all of today's integrated circuit devices.

Although virtually everyone involved with electronics has at least some experience with basic op amps, many hobbyists are not too familiar with some powerful variations on the basic operational amplifier. This book is intended to correct this situation.

Two important op amp variants are the transconductance operational amplifier, or OTA, and the Norton operational amplifier. The differences between the OTA and the Norton amplifier and the conventional op amp may seem a bit subtle at first, but they have some very important implications.

The conventional operational amplifier has two voltage inputs and a current output. The transconductance operational amplifier also has two voltage inputs and a current output. The Norton operational amplifier has two current inputs and a voltage output. All three devices can be used in many of the same applications, but the special features of the transconductance operational amplifier and the Norton operational amplifier make them suited to some extra uses of their own.

This book discusses the theory of operation for transconductance operational amplifiers and Norton operational amplifiers. Popular ICs of these types are presented in some detail. In addition, over 45 practical projects are included to illustrate some of the many possible applications, and to give you hands-on experience with these powerful variations on the basic operational amplifier.

The projects in this book cover a wide range: dc amplifiers, ac amplifiers, voltage and current switches, filters, voltage regulators, constant current sources, modulators, sample-and-hold circuits, Schmitt triggers, signal generators, temperature sensors, and a number of other devices.

All of these projects are fairly simple, and each could be constructed in a single evening. Using all new components, none of the projects in this book should cost more than $10 or $15.

Note: It is assumed that the reader has some prior background in electronics and project building. Although the function of the basic operational amplifier is briefly reviewed in this book, it would be helpful for the reader to already have some familarity with how this device works.

❖ Part I
Transconductance Operational Amplifiers

❖ 1
The Transconductance Operational Amplifier

MOST ELECTRONIC HOBBYISTS ARE FAMILIAR WITH THE STANDARD operational amplifier, or op amp. Many op amp ICs are available, from the very basic and inexpensive to some very deluxe units.

Operational amplifiers originally were designed primarily to perform mathematical operations in analog computation circuits. This intended function certainly explains the name.

Early op amp circuits, using discrete components, were complex, bulky and expensive, and so they were rarely used unless they were absolutely essential. With the development of the integrated circuit op amp in the late sixties and early seventies, this changed drastically. As costs and sizes dropped to almost negligible levels, countless new applications were found for the op amp. Today the operational amplifier IC is virtually considered an "all-purpose" device.

Sooner or later, specialized versions of all standard components start to show up, and the op amp is certainly no exception to this rule. One increasingly important variation on the basic operational amplifier is the *transconductance operational amplifier*, or *OTA*. For some reason, the standard word order is altered for the acronym—Operational Transconductance Amplifier.

WHAT IS AN OTA?

To oversimplify matters slightly, the term *transconductance* is roughly equivalent to the term *gain*. The gain of an amplifier is a measurement of the difference between the amplitudes of the input signal and the output signal.

Both a conventional operational amplifier and a transconductance operational amplifier have two inputs, inverting and noninverting, and one output. A transconductance operational amplifier also features an extra pin, which permits the circuit designer to control the gain, or transconductance of the amplifier externally and dynamically.

An input current, or *I-bias*, at this added pin is used to vary the transconductance of the device. The larger this input current is, the greater the gain of the amplifier.

An ordinary op amp roughly fits the model of a voltage source in series with an output resistance. A transconductance op amp, on the other hand, can be roughly visualized as a current source in parallel with an output resistance. This is a subtle, but important difference which defines the special features of the OTA.

In perhaps slightly oversimplified terms, an OTA is a current-controlled amplifier. The input (I-bias) current determines the operation (gain) of the amplifier.

In fact, you could go so far as to consider the transconductance control pin a programmable gain terminal, making the transconductance operational amplifier a programmable gain current-controlled amplifier.

HOW AN OP AMP WORKS

A standard operational amplifier is a voltage amplifying differential amplifier. The conventional schematic symbol for an op amp is shown in Fig. 1-1. Notice that it has two voltage inputs: V_1 (noninverting) and V_2 (inverting). The output is a voltage equal to:

$$V_o = G \times (V_1 - V_2)$$

where G is the gain of the device. The open loop (no feedback) gain of an op amp normally is extremely high. In most practical circuits, positive feedback is used to reduce the gain to a significantly lower level. This is done by setting up a feedback path between the op amp's output and its inverting input.

How an Op Amp Works 5

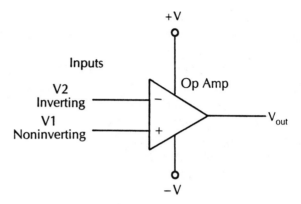

Fig. 1-1 *An op amp has two inputs and a single output.*

As an example, one of the most basic op amp circuits, the inverting amplifier, is illustrated in Fig. 1-2. Notice that there is only one signal input shown for this circuit. What happened to the second input? Nothing. The operational amplifier's noninverting input is simply grounded in this circuit. This is exactly the same as feeding in a constant input signal of 0 volts.

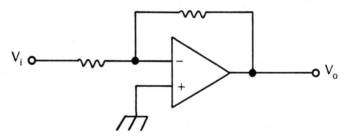

Fig. 1-2 *An inverting amplifier is one of the most basic op amp circuits.*

Since subtracting zero from any value results in the original value, we can simply ignore the second input in this application. The output voltage is controlled only by the signal at the inverting input and the circuit gain:

$$\begin{aligned} V_o &= G \times (V_1 - V_2) \\ &= G \times (0 - V_2) \\ &= G \times (-V_2) \end{aligned}$$

The negative sign indicates that the output signal is 180 degrees out of phase with the input signal. This is the effect of using the inverting input—to invert the signal polarity. If the input signal is positive, then the output signal will be negative, and vice versa.

Another basic op amp circuit is the noninverting amplifier circuit, illustrated in Fig. 1-3. In concept, this circuit is not too unlike the inverting amplifier shown in Fig. 1-2. Again, we are using only one of the operational amplifier's differential inputs, the noninverting input in this case. This circuit also employs negative feedback, by returning some of the output signal back to the inverting input to reduce the open loop again.

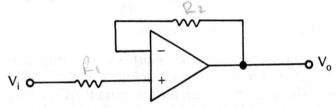

Fig. 1-3 Conventional op amps are also used in noninverting amplifier circuits.

Notice that the inverting input must be used for the negative feedback. The polarity inversion (phase shift) of the inverting input is an essential element of negative feedback. Using the noninverting input for the feedback would cause the feedback signal to add to the original input signal. The op amp would try to increase, rather than decrease, its open loop gain. At best, the result will be circuit instability or uncontrolled oscillations. In many cases, the op amp chip, and possibly some other components in the circuit could be damaged or destroyed.

The gain equation for a noninverting amplifier circuit is somewhat more complex than the one for an inverting amplifier. The noninverting gain equation is:

$$V_o = (1 + R_2/R_1) V_{in}$$

Notice that there is no negative sign in this equation because the noninverting input does not reverse the signal polarity. The output signal is in phase with the input signal.

Notice also that the gain of a noninverting amplifier circuit must always be at least equal to unity. Negative gain (attenuation) is not possible with this circuit.

The lowest possible gain for this circuit is achieved when both resistors are eliminated, as shown in Fig. 1-4. This effectively reduces the resistance values to zero. In this case, the gain works out to:

$$V_o = (1 + 0/0) \times V_{in}$$
$$= (1 + 0) \times V_{in}$$
$$= 1 \times V_{in}$$
$$= V_{in}$$

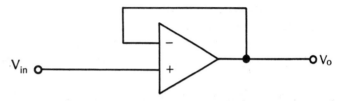

Fig. 1-4 *The lowest possible gain (unity) in a noninverting amplifier circuit is obtained when both resistors are omitted.*

Under these circumstances, we have unity gain. The output signal is identical to the input signal. Such a circuit is used as a buffer or an impedance matcher.

If both of the operational amplifier's inputs are used, we have a differential amplifier circuit, as illustrated in Fig. 1-5. Once again, negative feedback is used to reduce the open loop gain of the operational amplifier.

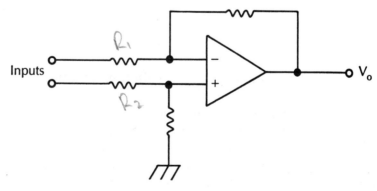

Fig. 1-5 *A differential amplifier circuit uses both of an op amp's inputs.*

In most cases, resistor R3 will be equal to resistor R4, and resistor R1 will be equal to resistor R2. These equalities don't always have to be true, but they do greatly simplify circuit design in many practical applications. There is rarely if ever any good reason for the circuit designer not to follow these resistance unities.

In any case, for a true differential amplifier, the R_3 to R_1 and R_4 to R_2 ratios must be equal. That is:

$$R_3/R_1 = R_4/R_2$$

The circuit will still function even if these ratio equalities are not maintained, but the signals at the inverting and noninverting inputs will be subjected to differing amounts of gain, which would be undesirable or inconvenient in most practical applications.

Generally speaking, it is easiest on the circuit designer to maintain the resistor value equalities just mentioned:

$$R_1 = R_2$$
$$R_3 = R_4$$

These resistor ratios determine the gain of the amplifier:

$$G = R_3/R_1 = R_4/R_2$$

Assuming the resistor ratio equalities are maintained, the output voltage will be equal to the difference between the two input voltages, multiplied by the gain. That is:

$$V_o = G \times (V_1 - V_2)$$

which is the general op amp gain equation given earlier in this section.

Of course a great many other operational amplifier circuits are possible, but most are, in one way or another, variants on the basic circuits presented here.

The input impedance of a standard op amp is typically very high, while the output impedance is generally quite low. Notice that in working with a conventional op amp circuit, we are only concerned with signals in the form of voltages.

HOW AN OTA WORKS

The schematic symbol for a transconductance operational amplifier is shown in Fig. 1-6. Compare this diagram to the symbol for a conventional op amp, shown in Fig. 1-1. Notice the two small circles, representing a constant current source at the output, and the extra control input, labeled "I_{bias}."

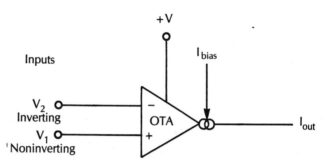

Fig. 1-6 *The schematic symbol for an OTA is similar to that for a conventional op amp.*

A standard op amp is a differential voltage amplifier. A transconductance operational amplifier also has a pair of differential (inverting and noninverting) inputs. In essence, this device is a differential voltage-to-current amplifier.

While the input signals are still assumed to be voltages, the output of an OTA is in the form of a high-impedance current. The output signal of a transconductance operational amplifier is defined as:

$$I_o = gm \times (V_1 - V_2)$$

where V_1 and V_2 are the differential input voltages, and I_o is the output current.

The term "gm" represents the transconductance, or current gain of the device. The gm value is typically given in mhos. A mho is just the opposite of an ohm, the measure of resistance. To convert between ohms and mhos, just take the reciprocal. That is,

$$X \text{ ohms} = 1/X \text{ mhos}$$
$$Y \text{ mhos} = 1/Y \text{ ohms}$$

The transconductance, voltage-to-current gain value, is always in direct proportion to the input current applied to the I-bias input of the OTA. In a practical transconductance operational amplifier IC, the acceptable range of values for the I-bias current runs from a low of 0.1 µA (microamperes) up to a high of 1 mA (milliampere). This might seem like a rather narrow range, but it permits the amplifier's gain to be varied over a 10,000:1 range.

In a practical device, the gm value is typically equal to twenty times the I-bias input value. This allows us to modify the output equation as follows:

$$I_o = 20\, I_{bias} \times (V_1 - V_2)$$

One major advantage of a transconductance operational amplifier is that it consumes very little power in most practical applications. Typically, the current consumed by the OTA IC is equal to two times the I-bias value setting the chip's transconductance.

Different applications will control the I-bias value in various ways. Probably the most common and the simplest way to set the I-bias value in a practical circuit is to use an external voltage source, passing the voltage through a series resistor to convert it to a current, according to Ohm's Law, at the I-bias input pin. Since the resistance is constant, Ohm's Law states that the current will vary proportionately to the voltage:

$$I = E/R$$

where I is the current, E is the voltage, and R is the resistance.

In most practical applications, a transconductance operational amplifier is used in the open loop mode. The gain is usually not limited with negative feedback, unless the OTA is being used to simulate a conventional op amp device. In most practical OTA circuits, the gain is controlled by the I-bias input.

OTA APPLICATIONS

As with the standard operational amplifier, the OTA has countless potential applications. Many of these applications are more or less shared with standard voltage type operational amplifier devices.

While the transconductance operational amplifier is a current controlled device, it can be used in voltage controlled circuits, such as VCAs (voltage controlled amplifiers), VCFs (voltage controlled filters), and VCOs (voltage controlled oscillators).

OTAs are also used for triangle-to-sine-wave conversion and in switching and Sample-and-Hold circuits. Other OTA applications include modulators and signal generators. Many practical circuits and applications for OTAs will be presented in later chapters of this book.

A transconductance operational amplifier can be employed in almost any standard, voltage, op amp application. This can be done simply by placing a suitable load resistor across the output, as shown in Fig. 1-7. An output voltage can then be tapped off across this resistor, thanks to Ohm's Law:

$$E = IR$$

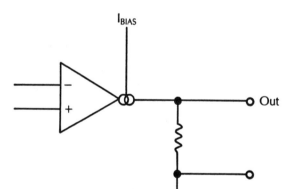

Fig. 1-7 An OTA's current output can be converted to a voltage by including a load resistor in the circuit.

where E is the voltage, I is the current, and R is the resistance. Since the resistance is a constant in this case, the output voltage, E, will vary proportionately with any changes in the output current, I.

12 The Transconductance Operational Amplifier

One reason to substitute an OTA in place of a conventional op amp is to reduce the power requirements of the circuit. In most circuits, the total amount of current consumed by an OTA is equal to twice the I-bias value. As stated earlier, I-bias can typically range from 0.1 μA up to 1 mA. Even at the upper end of the scale, the power consumption is obviously pretty miniscule.

❖2
The CA3080 and Other Typical OTA Devices

A NUMBER OF TRANSCONDUCTANCE OPERATIONAL AMPLIFIER ICs are now available to the circuit designer and the hobbyist. This chapter will take a look at a few typical OTA chips, and their specifications.

Probably the most popular and widely available OTA IC is the CA3080. The pin-out diagram for this device is shown in Fig. 2-1. Notice that only six of the eight pins are actually used on this chip. Pins 1 and 8 are not internally connected to anything. They are included solely to make up a standard size eight pin DIP package.

SPECIFICATIONS FOR THE CA3080

The CA3080 can be used with either a single-polarity or a dual-polarity power supply. A wide range of supply voltages may be used. For a single-polarity power supply, the minimum supply voltage is +4 volts dc, and the maximum supply voltage is +36 volts dc. Dual-polarity power supplies may be used over a similar range with the CA3080. The limits are ±2 volts dc to ±18 volts dc. Supply voltages of ±9 volts to ±15 volts are recommended for the best possible and most reliable results in most practical applications.

The CA3080 can safely dissipate up to 125 milliwatts (0.125 watt). The differential input voltage (the voltage difference between the inverting and noninverting inputs) is affected by the chip's supply voltage. Assuming the maximum supply voltage is being used, the maximum differential input voltage is ±18 volts. The maximum differential input voltage must never exceed the circuit's supply voltage.

The CA3080 features a typical Common Mode Rejection Ratio of 110 dB. That is, if identical signals are simultaneously fed to the two differential inputs, inverting and noninverting, the signal will be suppressed at the output by a factor of approximately 110 decibels.

The CA3080's bandwidth when used in the open loop mode is 2 MHz (2,000,000 Hz). Of course, when a feedback path is used to reduce the open loop gain, the frequency response bandwidth may be reduced somewhat, depending on the specific feedback components being used in the circuit.

The input signal current is internally limited to a maximum value of 1 mA (0.001 ampere). The amplifier bias current (I-bias) must not exceed 2 mA (0.002 ampere).

A very nice feature of the CA3080, which is different from most conventional op amp devices, is that the output can theoretically be short-circuited indefinitely without damaging the chip. Of course, depending on the specific design of the circuit, other components could quickly be damaged by such a short circuit. The CA3080's design only protects itself.

If the CA3080 is used in a unity gain circuit, the slew rate is 50 volts-per-microsecond. This value will change for circuits with nonunity gains.

An unusual specification for transconductance operational amplifiers is the forward transconductance rating. For the CA3080 the typical forward transconductance is listed on the specification sheet as 9600 μmho (micromhos). There will be some variation from unit to unit.

Many of the CA3080's operating parameters are directly related to the I-bias input value. This current controls much of the operation of the IC.

The I-bias current sets the transconductance (gm gain) of the OTA. This is the primary function of the I-bias input. For example, the maximum output current for the CA3080 is equal to the value of I-bias. There is no way to get an output current higher than the current I-bias value from this OTA chip.

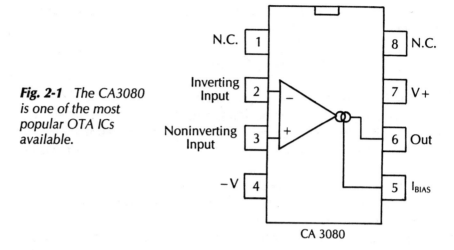

Fig. 2-1 The CA3080 is one of the most popular OTA ICs available.

The operating current drawn by the CA3080 is also controlled by the bias input current. The current drawn by the chip at any given instant is equal to twice the present I-bias value.

The input I-bias current drawn by the inverting (pin 2) and noninverting (pin 3) inputs is also affected by the I-bias value. The exact value will depend somewhat on the current gain of the transistors in the on-chip differential amplifier circuit, and it will fluctuate somewhat from device to device. As a rule, the current drawn by the op amp inputs will be approximately equal to I-bias/200 per input.

The CA3080's input and output impedances are also affected by the I-bias current value. The decrease of both the input and output impedances is linear with increases in the transconductance. The output impedance is always greater than the input impedance, generally by a factor of approximately ten.

The range of available output voltages is determined by the I-bias value and an external load resistor which is connected to the chip's output (pin 6). The peak output voltage swing is found simply by taking the product of these two values:

$$I_{bias} \times R_1$$

As an example, let's assume that the external load resistor has a value of 100K (100,000 ohms), and the current at the I-bias input is 10 µA (0.00001 ampere). In this case the peak output voltage swing works out to:

$$0.00001 \times 100000 = 1 \text{ volt}$$

There is one special case to consider. What if the load impedance is infinite (no external load resistor)? Under this circumstance, the output voltage can reach a maximum value 1.5 volts below the circuit's positive supply voltage, and a minimum value 0.5 volt above the circuit's negative supply voltage.

The OTA's bandwidth and slew rate are also influenced by the I-bias value and the external load capacitance, if any, connected to the pin 6 output.

The slew rate is a measure of how fast the output signal can change in response to rapid changes in the input signal. This specification is normally given in terms of volts-per-microsecond, or v/µS.

If the circuit includes an external load capacitance (C_1), the slew rate will be defined as:

$$I_{bias}/C_1$$

where C_1 is the capacitance in picofarads (pF). In circuits without an external load capacitor, the maximum possible slew rate for the CA3080 is approximately 50 V/µS.

For the most reliable results, the I-bias current should not be permitted to drop below about 0.5 µA (microampere), or rise above approximately 0.5 mA (milliampere). The upper limit is especially important. If too large an I-bias current is fed to the CA3080, the chip could overheat and go into thermal runaway. This could quickly damage or destroy the IC, and possibly some other components in the circuit.

THE CA3080'S INTERNAL STRUCTURE

Figure 2-2 shows a simplified schematic for the internal circuitry of the CA3080 transconductance operational amplifier IC.

Breaking the CA3080 up into functional subcircuits, we find that this device consists of a differential amplifier and four special subcircuits known as *current mirrors* (CMs).

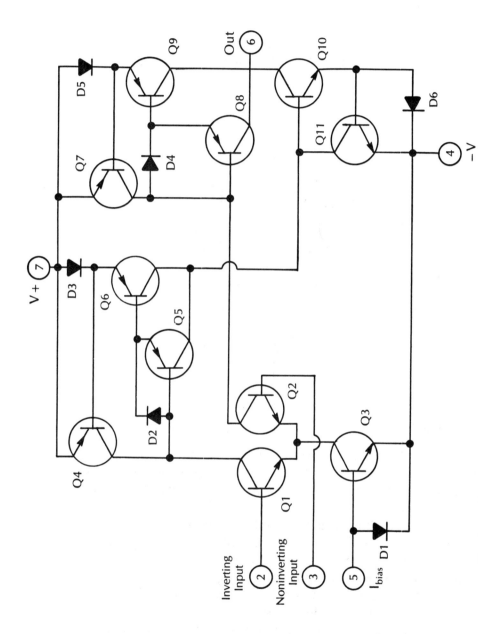

Fig. 2-2 This is a simplified diagram of the internal structure of a CA3080 OTA IC.

18 The CA3080 and Other Typical OTA Devices

A very simplified version of a CA3080's differential amplifier circuit is shown in Fig. 2-3. The key equations for this circuit are as follows:

$$I_c = I_a + I_b$$
$$I_b - I_a = V_{in} \times gm = 20 V_{in} \times I_c$$

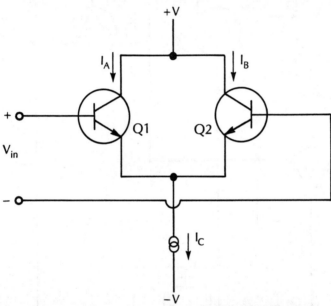

Fig. 2-3 *A CA3080 OTA includes a differential amplifier.*

Actually, 19.2 $V_{in} \times I_c$ would be more accurate, but 20 is a more convenient value than 19.2, and the introduced error from rounding up will be insignificant in the majority of practical applications.

In these equations, I_c is the combined emitter current of the two transistors. I_a is the collector current of the first transistor, while I_b is the collector current of the second transistor. V_{in}, of course, is the differential input voltage applied between the circuit's two signal inputs—inverting (−) and noninverting (+). Finally, gm is the transconductance.

If identical signals are fed to both the inverting input and the noninverting input, then the differential input voltage (V_{in}) will be zero. Under these circumstances, the two collector currents, I_a

and I_b will take on equal values. Since the emitter current (I_c) is the sum of the two collector currents, $I_a + I_b$, it follows that either of the collector currents is equal to one half of the emitter current:

$$I_a = I_c/2$$
$$I_b = I_c/2$$

If the signals at the differential amplifier's inverting and noninverting inputs are not equal, then the differential input voltage (V_{in}) will obviously have a nonzero value. For this particular circuit, the differential input voltage should be limited to a maximum value of ±25 mV (0.025 volt). If V_{in} is not equal to zero, then the two collector currents, I_a from I_b, will take on different values. Subtracting I_a from I_b gives us the equivalent of the differential input voltage, V_{in}, multiplied by the transconductance (gm).

The value of gm is at all times directly proportional to the emitter current (I_c). Nominally:

$$gm = 20 \times I_c$$

The constant in this equation may vary with changes in temperature. The value of 20 is assuming an ambient temperature of approximately 25°C.

This differential amplifier circuit (Fig. 2-3) is pretty unimpressive by itself. The power of the transconductance operational amplifier lies in the addition of the current mirror subcircuits. A current mirror can be used as a sink, as shown in Fig. 2-4, or as a

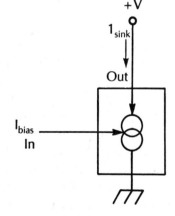

Fig. 2-4 A current mirror can be used as a current sink.

source, as shown in Fig. 2-5. When used as a sink, the current mirror will sink as much current as is applied to its input. Or, a current mirror source will supply as much current to its output as is applied to its input.

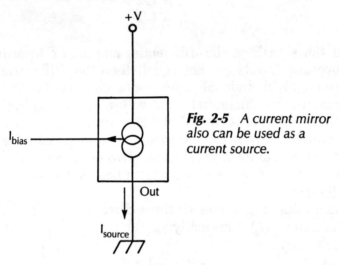

Fig. 2-5 A current mirror also can be used as a current source.

Using a current mirror with the simple differential amplifier circuit of Fig. 2-3 permits external control of the emitter current (I_e), and therefore of the transconductance of the amplifier.

Three more current mirror circuits are included in the CA3080 chip to make the difference current, $I_b - I_a$, externally available, which is a very useful property in many practical circuit designs.

A simplified current mirror circuit is illustrated in Fig. 2-6. As you can see, this is a relatively simple circuit. This current mirror circuit is designed for use as a sink. There are some minor differences in a current mirror source circuit.

Since its collector is shorted to its base, transistor Q1 effectively acts as a diode. This pseudo-diode is wired across the base-emitter junction of a second, closely matched transistor. The transistors must be identical in order for the circuit to function properly. The two transistor mirror approach is taken to make the circuit less sensitive to the current gains of the transistors themselves. The output impedance is also significantly increased by using this particular design approach.

In two of the CA3080's internal current mirror circuits, Darlington transistor pairs and speed-up diodes are included to

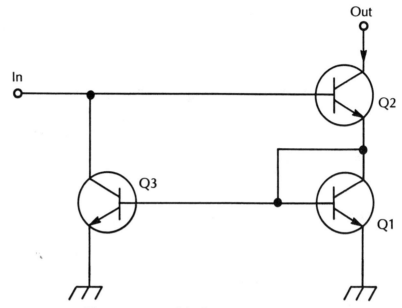

Fig. 2-6 *A simplified current mirror circuit*

improve the circuit performance. These refinements are not illustrated in Fig. 2-6 to make the basic principles involved clear. The Darlington transistor pairs and the speed-up diodes are shown in Fig. 2-2.

Figure 2-7 is a simplified block diagram of the complete circuitry in a CA3080 transconductance operational amplifier IC.

The bias current (I-bias) input signal is fed through current mirror C to control the emitter current; therefore the emitter current corresponds to the transconductance of the differential amplifier.

Collector current I_a (transistor Q1) is mirrored by current mirror A, while current mirror B does the same for collector current I_b (transistor Q2). These two currents are fed to current mirror D. The difference value of the two collector currents, $I_b - I_a$, can be tapped off between current mirror B and current mirror D. This is the output of the OTA (pin 6 on the CA3080).

COMPENSATION

The CA3080 transconductance operational amplifier is internally uncompensated. In most applications, this device will be used in an open-loop configuration, so external compensation

22   The CA3080 and Other Typical OTA Devices

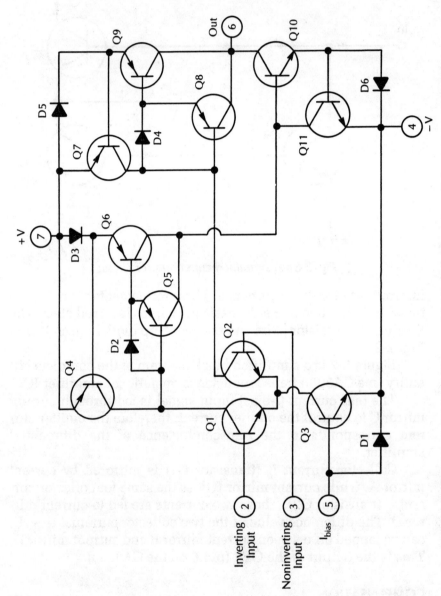

Fig. 2-7 This block diagram illustrates the internal circuitry of the CA3080 OTA IC.

is rarely required in the circuit design. As a general rule, external frequency compensation is required only in circuits using negative feedback with the OTA.

If you have done much work with conventional operational amplifiers, you should already be familiar with the concept of frequency compensation. Many op amp ICs, such as the 741, are internally compensated, but others require an external frequency compensation network, usually just a simple capacitor or an RC (resistor-capacitor) network.

When negative feedback is used with an op amp or an OTA, some of the output signal is fed back to the amplifier's inverting input. At relatively low frequencies, this feedback signal is, by definition, 180 degrees out of phase with the original input signal. This negative feedback causes some of the input signal to be cancelled, effectively limiting the gain of the amplifier.

Unfortunately, as the input frequency increases, so does the phase shifting within the op amp circuitry. If the signal frequency is made high enough to produce another 180 degrees of phase shift, the difference between the input and output signals will now be 360 degrees, or one full cycle. In effect, the input and output signals will be in phase with one another. They will now add, instead of partially cancelling each other. If the closed-loop gain is greater than 1 (unity), this can lead to all sorts of problems. The op amp can be forced into saturation and severe clipping of the output signal. In addition, instability or oscillation can result.

To compensate for such problems, a simple RC network, like the one shown in Fig. 2-8, is used to control the maximum phase shift of the op amp. Many practical op amp ICs, like the 741, include the frequency compensating RC network within the

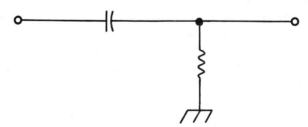

Fig. 2-8 *Frequency compensation can be achieved by adding a simple RC network to the circuit.*

chip's internal circuitry. Others have a special set of pins, permitting the circuit designer to add an external frequency compensation network to suit the intended use.

Since the CA3080 and other OTAs are generally used in open loop circuits, frequency compensation is usually not a problem. Even in applications involving negative feedback, frequency compensation is not necessary if all signal frequencies are relatively low. The CA3080 is rarely used in high-frequency, negative feedback circuits. However, you should be aware of the compensation issue, just in case you happen to come across an unusual circuit with compensation problems.

THE LM13600 DUAL OTA

Another popular transconductance operational amplifier IC is the LM13600. Two complete and independent OTAs are contained within a single 16 pin DIP housing, along with two separate buffers. The only pins used in common by both of the OTAs are the power supply terminals.

The pin-out diagram for the LM13600 appears as Fig. 2-9. Either a single end or a dual polarity power supply may be used with this chip. If a single polarity voltage source is used, the supply voltage may be as large as +36 volts dc. For a dual polarity power supply, the maximum supply voltage is ±18 volts.

The two internal OTA stages are fully independent, except for the power supply, but being fabricated on a single chip, they are very closely matched, which can be highly desirable in certain precision uses. In a typical device, the transconductance values are matched within 0.3 dB. This close matching between the amplifiers makes the LM13600 an excellent choice for high-fidelity stereo applications.

The OTAs themselves feature a high-impedance output. In some applications, it may be preferable to have a low-impedance output. In such a case, some sort of buffer amplifier is needed. The LM13600 provides a suitable buffer stage for each OTA on-chip. Use of the buffer is entirely optional, depending on the design of the circuit.

Each buffer is a Darlington transistor pair, as illustrated in Fig. 2-10. Because of the base-emitter voltage drops of the buffer transistors, the buffered output voltage will be about 1.2 volts

The LM13600 Dual OTA   25

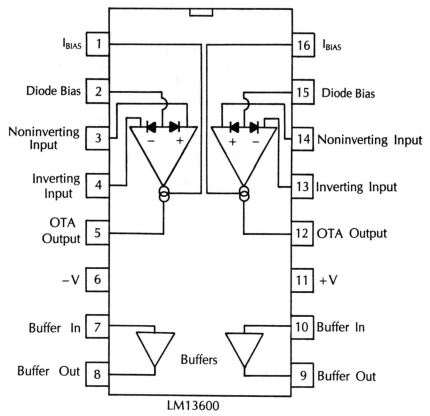

Fig. 2-9 *The LM13600 contains two independent OTAs in a single package.*

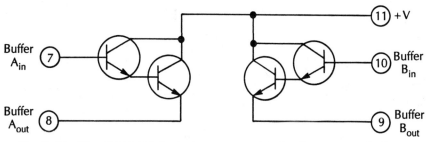

Fig. 2-10 *The LM13600 includes two on-chip Darlington pair buffers.*

lower than the straight (unbuffered) OTA output. This could be a problem in some precision dc amplifier applications.

If the direct unbuffered output signal is desired, it can simply be tapped off from pin 5 or pin 12. If the buffer stage is to be included in the circuit, the OTA output, pin 5 or 12, is connected to the appropriate buffer input, pin 7 or 10, and the circuit's output signal is tapped off from the appropriate buffer output, pin 8 or 9. The buffer output, pin 8 or 9, should be connected to the negative supply voltage, or ground, if a single-polarity power supply is being used, through a suitable load resistor.

Because the buffers are not internally connected to the OTAs, they can be used for other purposes in some circuits. External current limiting is advisable for the buffer outputs. No more than about 20 mA (0.02 ampere) should be drawn from the buffer transistors.

A very simplified circuit diagram of one of the LM13600's transconductance operational amplifiers appears in Fig. 2-11. A more detailed schematic is shown in Fig. 2-12.

Each OTA in the LM13600 is quite similar to the CA3080, discussed earlier in this chapter, with one significant difference.

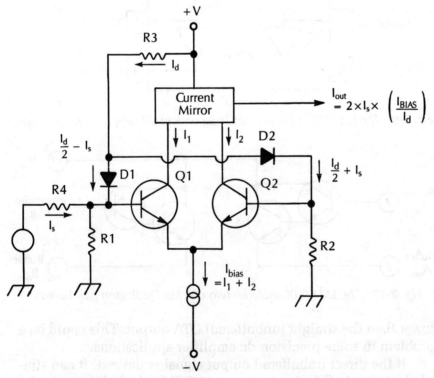

Fig. 2-11 *A simplified version of the OTA circuit contained in the LM13600.*

The LM13600 Dual OTA 27

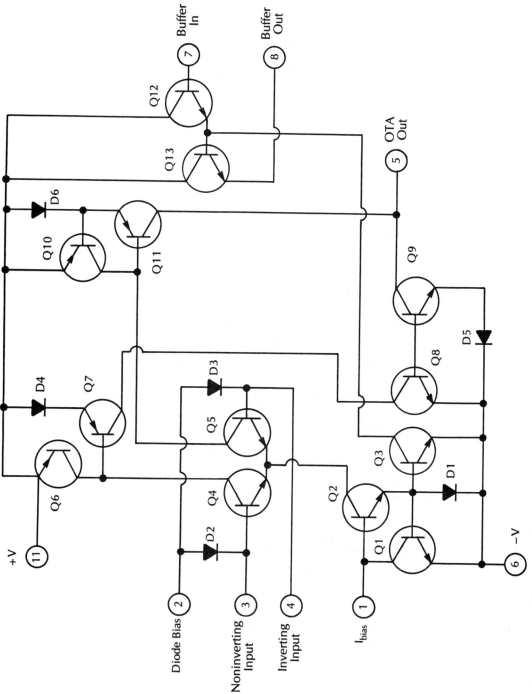

Fig. 2-12 This diagram offers greater detail in the OTA circuits of the LM13600 chip.

The LM13600's OTAs feature a pair of linearizing diodes, as illustrated in Fig. 2-13. These diodes are shown as D1 and D2 in Fig. 2-11. The diodes are specifically designed to have characteristics that are closely matched to those of the base-emitter junctions of the transistors, Q1 and Q2.

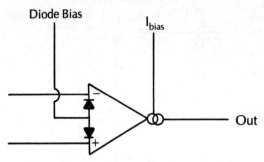

Fig. 2-13 *The LM13600's OTAs feature a pair of linearizing diodes between the inputs.*

In a practical circuit, resistors R1 and R2 are wired between the differential inputs and ground, as shown in Fig. 2-11. These resistors have equal, fairly low values. Typically, 470 ohms or so is the best value for these input resistors.

A bias current (I_d) from the positive supply voltage through resistor R3 is fed to the input resistors via the linearizing diodes. The diodes are matched, having identical characteristics, and the input resistors have equal values. As a result, the bias current, I_d, is equally divided between the two input resistors.

The input signal voltage is applied to the circuit through resistor R4. This resistor's value is relatively large compared to R1 and R2.

The input voltage across resistor R4 creates an input signal current, I_s, which feeds into resistor R1. A voltage is generated across R1, reducing the current through diode D1 to a value of:

$$(I_d/2) - I_s$$

The bias current, I_d, is a constant value, so the current through diode D2 is forced to increase accordingly, to a value of:

$$(I_d/2) + I_s$$

In effect, the linearizing diodes create a large negative feedback to the differential amplifier. This feedback works to minimize the signal distortion.

The gain, or transconductance of the OTA can be controlled by changing the value of I_d, by changing the value of resistor R3, or by adjusting the control input current (I-bias). Both I_d and I-bias should be limited to 2 mA or less. Assuming that the input signal current, I_s, is relatively small compared to the bias current, I_d, the circuit's output current is equal to:

$$I_o = 2I_s(I_{bias}/I_d)$$

By simply ignoring the linearizing diode bias input, pins 2 and 15, the LM13600 can be used as a substitute for a "standard" transconductance operational amplifier, such as the CA3080.

In the following chapters, we will present and examine a number of practical circuits and uses for OTAs, such as the CA3080 and the LM13600.

❖3
Amplifier Projects

CHAPTERS 3 THROUGH 6 WILL PRESENT A NUMBER OF PRACTICAL transconductance operational amplifier projects for you to build and experiment with. There are many possible applications for OTAs. This chapter deals with a rather obvious category—amplifiers.

In very simple terms, an amplifier is usually a circuit which accepts an input signal and produces an output signal which is identical to the input signal except for a change in amplitude.

Most amplifier circuits increase the amplitude of the input signal; that is, the gain is positive. A few amplifier circuits, however, reduce the amplitude of the input signal with a negative gain. Such circuits are often called *attenuators*.

There is one special class of amplifier circuits: when the amplitude of the output signal is the same as the amplitude of the input signal, the gain is one, or unity. An amplifier with a gain of one is usually referred to as a buffer. Buffers have a number of uses, primarily impedance matching, load isolation, and polarity inversion.

A transconductance operational amplifier, like a conventional op amp, can be used in three basic types of amplifier circuits. They are:

- inverting amplifier
- noninverting amplifier
- differential amplifier

In an inverting amplifier circuit, only a single input signal is used. This input signal is fed into the OTA's inverting input. The noninverting input is grounded usually, but not always, through a resistor, so its effective input voltage is zero. The output signal's polarity is inverted from the input signal's. That is, if the input signal is positive, then the output signal is negative, and vice versa. For ac signals, the output signal is phase shifted 180 degrees.

A noninverting amplifier circuit also uses only a single input signal, but in this case the input signal is fed into the OTA's noninverting input. The inverting input is grounded usually, but not always, through a resistor, so its effective input voltage is zero. The output signal's polarity is the same as that of the input signal. That is, if the input signal is positive, then the output signal is also positive, and vice versa. For ac signals, the output signal is in phase with the input signal.

A differential amplifier circuit uses both of the OTA's inputs. Usually, two separate input signals will be fed into the circuit. The signal voltage at the inverting input will be subtracted from the signal voltage at the noninverting input, and the difference between these two signals will be amplified by the gain, or transconductance, of the circuit.

You will be able to see that all of the amplifier circuits presented in this chapter are variations of one or another of these three basic amplifier types.

DIRECT-COUPLED DIFFERENTIAL AMPLIFIER

Figure 3-1 shows one of the most basic circuits for the CA3080 transconductance operational amplifier. Both of the OTA's inputs are being used here, so this is obviously a difference amplifier. A typical parts list for this project appears in Table 3-1. Alternative component values may be substituted to suit your individual application. Try breadboarding this circuit and experimenting with various resistor values.

The input voltages to this circuit are fed directly into the OTA through a pair of resistors (R1 and R2). This is known as *direct coupling*.

These input resistors have equal values to keep the input signals balanced and equally weighted. In other words, the input resistors help maintain the dc balance of the OTA. These resistors also aid in equalizing the source impedances of the two inputs.

The gain of this amplifier is dependent on the value of the I-bias current. As you will see in the following discussion, the I-bias current is in turn dependent on the supply voltage used for the circuit.

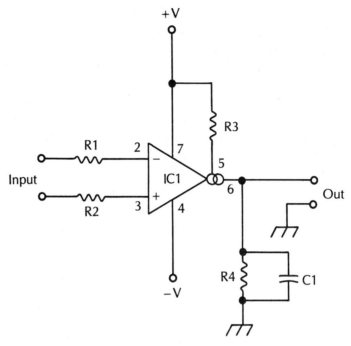

Fig. 3-1 Direct-coupled Differential Amplifier

**Table 3-1 Parts List for the
Direct-coupled Differential Amplifier project of Fig. 3-1.**

Part	Description
IC1	CA3080 OTA IC
C1	200 pF capacitor
R1, R2	2.2K 5 percent 1/4-watt resistor
R3	33K 5 percent 1/4-watt resistor
R4	10K 5 percent 1/4-watt resistor

The I-bias terminal, pin 5, on the CA3080 OTA chip is internally connected to the negative supply voltage pin (4) through a chip transistor's base-emitter junction. Because of this internal connection, the biased voltage at pin 5 (I-bias input) is always approximately 600 mV (0.6 volt) more positive than the chip's negative supply voltage (V – at pin 4).

The I-bias input is usually connected to the circuit's positive supply voltage through a suitable current-limiting resistor. The necessary value of this resistor will depend on the intended use.

The value of resistor R3 sets the value of the I-bias current, and therefore the transconductance, or gain, of the amplifier. This resistor could be made variable to permit manual adjustment of the circuit's gain, if such a modification suits your purpose. For now, however, we will assume that a fixed resistor is being used for R3.

Resistor R3 forms a voltage divider with the internal base-emitter junction between pins 5 and 4. This voltage divider is placed across the positive and negative supply voltages, as illustrated in Fig. 3-2.

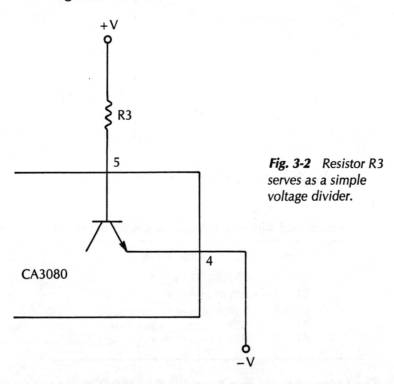

Fig. 3-2 Resistor R3 serves as a simple voltage divider.

To determine the value of the I-bias current, we first need to know the values of the circuit's supply voltages and of resistor R3. For our discussion, we will assume a ±9-volt dual-polarity power supply is being used to operate the circuit, giving a total effective supply voltage of 18 volts. We will also assume that resistor R3 has the value suggested in the parts list—33K (33,000 ohms).

Since we know the total voltage is 18 volts, and the voltage drop between pins 5 and 4 is 0.6 volt, it is obvious that the voltage dropped across resistor R3 must be about 17.4 volts (18 − 0.6). Now we can use Ohm's Law to determine the current through this resistor:

$$\begin{aligned} I &= E/R \\ &= 17.4/33000 \\ &= 0.0005 \text{ ampere} \\ &= 0.5 \text{ mA} \end{aligned}$$

The results of this equation have been rounded off for convenience.

This is the current seen by the CA3080's pin 5, so the value of I-bias is approximately 0.5 mA (milliampere). The total current drawn by the IC is about twice the I-bias value, or 2 × 0.5 = 1 mA, in this case.

By placing a load resistor (R4) across the circuit's output, between the IC's pin 6 and ground, the OTA's output current can be converted into an output voltage.

With an I-bias value of 0.5 mA, the transconductance of the CA3080 OTA is about 10 millimhos (0.01 mho). The gain of the amplifier circuit is the product of the OTA's transconductance and the load resistance, which is resistor R4 in this circuit. Again assuming that the suggested parts list of Table 3-1 is being followed, the value of the load resistor, R4, is 10K (10,000 ohms). This means that the gain of this circuit works out to:

$$\begin{aligned} \text{GAIN} &= gm \times R_4 \\ &= 10 \text{ mmhos} \times 10000 \text{ ohms} \\ &= 100 \\ &= 40 \text{ dB} \end{aligned}$$

For many practical applications, it is useful to convert the gain factor into decibels (dB).

Because of the internal design of the CA3080 chip, the maximum current that can flow into the load resistor is equal to the I-bias current, or 0.5 mA in our sample design. Using Ohm's Law again, this gives us a peak output voltage equal to:

$$E = IR$$
$$= 0.0005 \times 10000$$
$$= 5 \text{ volts}$$

The value of the load resistor, R4, also sets the output impedance of the amplifier circuit. The output impedance is simply equal to the value of resistor R4, or 10K in our sample circuit.

Capacitor C1 is also part of the load as seen by the OTA. This capacitor's value controls the slew rate limit of the circuit according to this formula:

$$I_{bias}/C_1$$

Plugging in the values from our design example, we get a slew rate approximately equal to:

$$0.5 \text{ mA}/200 \text{ pF}$$
$$= 2.5 \text{ V}/\mu S$$

or 2.5 volts per microsecond.

Notice that the OTA is being used here in an open-loop configuration. No feedback is employed in this circuit. If capacitor C1 was not included to limit the amplifier's slew rate, the CA3080 would operate at its maximum slew rate and bandwidth. The first reaction to this might be, "Well, what's wrong with that? The wider the bandwidth and the faster the slew rate, the better the circuit. Right?" Not necessarily. An excessive bandwidth or slew rate can cause instability problems, and the CA3080 might become excessively noisy. This would plainly be undesirable if the application were even moderately critical.

AC-COUPLED INVERTING AMPLIFIER

The amplifier circuit shown in Fig. 3-3 bears many similarities to the differential amplifier circuit of Fig. 3-1. There are also some key differences.

The most obvious difference between the two circuits is that the circuit of Fig. 3-3 uses just one of the OTA's inputs. The

AC-Coupled Inverting Amplifier

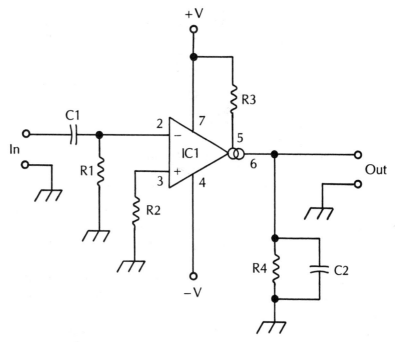

Fig. 3-3 AC-coupled Inverting Amplifier

inverting input is the one used to accept the external input voltage, and the noninverting input is grounded through resistor R2. This means this circuit is an inverting amplifier.

Another important difference here is the addition of the input capacitor, C1. This capacitor blocks any dc component in the input signal. Only ac signals can pass through a capacitor. The exact value of this capacitor is not critical. This circuit design technique is known as ac coupling.

A typical parts list for this ac coupled inverting amplifier circuit is given in Table 3-2. Try breadboarding this circuit and experimenting with various resistor values. Notice that the two input resistors, R1 and R2, have equal values to balance the OTA.

The gain of this amplifier, as with a transconductance operational amplifier, is dependent on the value of the I-bias current, which is in turn dependent on the supply voltage used for the circuit.

The I-bias terminal (pin 5) on the CA3080 OTA chip has an internal voltage drop of about 0.6 volt over the chip's negative supply voltage (pin 4).

Table 3-2 Parts List for the AC-coupled Inverting Amplifier project of Fig. 3-3.

Part	Description
IC1	CA3080 OTA IC
C1	0.5µF capacitor
C2	200 pF capacitor
R1, R2	15K 5 percent ¼-watt resistor
R3	33K 5 percent ¼-watt resistor
R4	10K 5 percent ¼-watt resistor

The I-bias input is connected to the circuit's positive supply voltage through a suitable current-limiting resistor. The necessary value of this resistor will depend on the specific application intended.

The value of resistor R3 sets the value of the I-bias current, and therefore the transconductance, or gain of the amplifier. This resistor could be made variable to permit manual adjustment of the circuit's gain, if such a modification suits your individual need. For now however, we will assume that a fixed resistor is being used for R3.

Resistor R3 forms a voltage divider with the internal base-emitter junction between pins 5 and 4. This voltage divider is placed across the positive and negative supply voltages.

To determine the value of the I-bias current, we first need to know the values of the circuit's supply voltages and of resistor R3. For our discussion, we will assume a ±9-volt dual-polarity power supply is being used to operate the circuit, giving an absolute value of 18 volts. We will also assume that resistor R3 has the value suggested in the parts list—33K (33,000 ohms).

Since we know the total voltage is 18 volts, and the voltage drop between pins 5 and 4 is 0.6 volt, it is obvious that the voltage dropped across resistor R3 must be about 17.4 volts (18 − 0.6). Now we can use Ohm's Law to determine the current through this resistor:

$$\begin{aligned} I &= E/R \\ &= 17.4/33000 \\ &= 0.0005 \text{ ampere} \\ &= 0.5 \text{ mA} \end{aligned}$$

The results of this equation are rounded off for convenience.

This is the current seen by the CA3080's pin 5, so the value of I-bias is approximately 0.5 mA (milliampere). The total current drawn by the IC is about twice the I-bias value, or 2 × 0.5 = 1 mA, in this case.

By placing a load resistor, R4, across the circuit's output (between the IC's pin 6 and ground), the OTA's output current can be converted into an output voltage.

With an I-bias value of 0.5 mA, the transconductance of the CA3080 OTA is about 10 millimhos (0.01 mho). The gain of the amplifier circuit is the product of the OTA's transconductance and the load resistance, which is resistor R4 in this circuit. In our sample design, using the parts values listed in Table 3-2, the gain of this circuit works out to about 100, or 40 dB.

Because of the internal design of the CA3080 chip, the maximum current that can flow into the load resistor is equal to the I-bias current, or 0.5 mA in our sample design. Using Ohm's Law again, this gives us a peak output voltage equal to:

$$E = IR$$
$$= 0.0005 \times 10000$$
$$= 5 \text{ volts}$$

The value of the load resistor, R4, also sets the output impedance of the amplifier circuit. The output impedance is simply equal to the value of resistor R4, or 10K in our sample circuit.

Capacitor C1 is also part of the load as seen by the OTA. This capacitor's value controls the slew rate limit of the circuit:

$$I_{bias}/C_1$$
$$0.5 \text{ mA}/200 \text{ pF}$$
$$= 2.5 \text{ V}/\mu S$$

or 2.5 volts per microsecond.

Notice that the OTA is being used here in an open-loop configuration. No feedback is employed in this circuit. If capacitor C1 was not included to limit the amplifier's slew rate, the CA3080 would operate at its maximum slew rate and bandwidth. An excessive bandwidth or slew rate can cause instability problems, and the CA3080 might become excessively noisy. This would plainly be undesirable if the application were even moderately critical.

LOW-POWER INVERTING AMPLIFIER

One of the major advantages of a transconductance operational amplifier is its relatively low power consumption. The circuits of Figs. 3-1 and 3-3 draw only about 1 milliamp from the power supply, which certainly isn't much. However, current drain can quickly add up in large multistage systems. This can be a particular problem in applications using battery power where current drain must be kept to an absolute minimum.

One solution is to use the CA3080 in a negative feedback circuit, almost as if it were a conventional op amp device. By doing this, the voltage gain of the amplifier is made nearly independent of the I-bias and supply voltage values.

A very low-power inverting amplifier circuit is shown in Fig. 3-4. A suitable parts list for this project appears in Table 3-3.

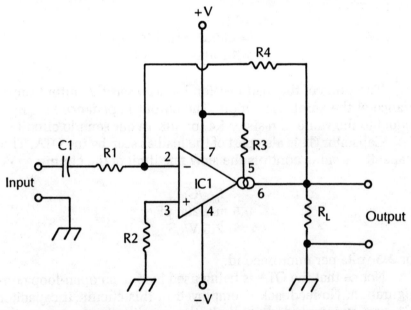

Fig. 3-4 Low-power Inverting Amplifier

Using a 9-volt dual-polarity power supply and the component values given in the suggested parts list, the amplifier gain is approximately 20 dB and the circuit's current drain is a mere 150

**Table 3-3 Parts List for the
Low-power Inverting Amplifier project of Fig. 3-4.**

Part	Description
IC1	CA3080 OTA IC
C1	0.5 µF capacitor
R1, R2	12K 5 percent ¼-watt resistor
R3	330K 5 percent ¼-watt resistor
R4	120K 5 percent ¼-watt resistor

µA (0.00015 ampere). The power consumed by this circuit is almost negligible.

For the time being, we will assume that the external load impedance (R_1) is infinite. In this case, the amplifier's voltage gain is set primarily by the ratio of the values of resistors R4 and R1, just as an ordinary inverting amplifier circuit built around a standard op amp:

$$A_v = -R_4/R_1$$

The negative sign, of course, indicates the polarity inversion resulting from the use of the inverting input.

Using the resistor values suggested in the parts list, the nominal gain works out as follows:

$$R_1 = 12K$$
$$R_4 = 120K$$
$$A_v = -120000/12000$$
$$= -10$$
$$= -20 \text{ dB}$$

This simple gain equation is truly valid only when the load impedance (R_1) is infinite. This restriction is due to the fact that the circuit's output impedance is set by the value of resistor R4 and the voltage gain:

$$Z_o = R_4/A_v$$
$$= 120000/10$$
$$= 12000 \text{ ohms}$$
$$= 12K$$

Any external (finite) load is effectively in parallel with the circuit's output impedance and thus lessens the output impedance value, reducing the output of the amplifier.

Even though conventional op amp techniques are used in this circuit, the OTA's I-bias pin is still being employed. The primary function of the I-bias value in this particular use is to set the circuit's overall operating current and limit the maximum amount of output swing.

Using the component values given in Table 3-3, mainly resistor R3 in this case, the value of I-bias is fixed at approximately 50 μA (0.00005 ampere). The total amount of current consumed by the circuit is about 150 μA (0.00015 ampere), as stated earlier in this section.

The maximum output voltage swing will be controlled by both the I-bias value and the load and circuit output impedances. Ohm's Law can be used to determine the maximum output voltage from the circuit from the values of I-bias and feedback resistor R4:

$$E = I_{bias} \times R_4$$

Remember, we are assuming that the component values from the parts list are being used throughout our discussion.

$$E = 0.00005 \times 120000$$
$$= 6 \text{ volts}$$

This simple equation assumes that the load impedance (R_1) is infinite. When there is a practical (finite) load impedance at the output of this circuit, the amplifier's output is reduced proportionately.

By selecting suitable resistor values in the circuit and an appropriate load impedance, the circuit designer can set up a wide range of possible voltage gains and peak output values, to suit a variety of practical uses.

Notice that since the OTA is being used in the closed loop mode in this circuit, an external slew rate limiting capacitor is not needed here.

VARIABLE-GAIN INVERTING AMPLIFIER

So far, the amplifier circuits presented in this chapter have featured fixed gains, set directly by the component values within each individual circuit. In some applications, manually adjustable gain will be desirable, or perhaps even essential.

The ac amplifier circuit shown in Fig. 3-5 permits the circuit operator to manually control the amplifier gain via potentiometer R6. Notice that this is basically another inverting amplifier circuit. A suggested parts list for this project is given in Table 3-4.

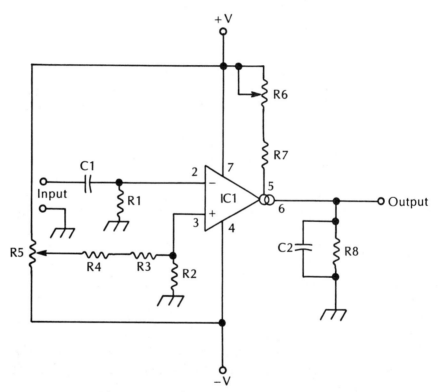

Fig. 3-5 *Variable-gain Inverting Amplifier*

Resistors R1 through R5 are used to bias the OTA's inputs. Potentiometer R5 is used to manually adjust the input bias. With a zero volt input signal, adjust this potentiometer for a true zero output. In most practical applications, it would make sense to

Table 3-4 Parts List for the Variable-Gain Inverting Amplifier project of Fig. 3-5.

Part	Description
IC1	CA3080 OTA
C1	0.5 µF capacitor
C2	200 pF capacitor
R1, R2	4.7K 1/4-watt 5 percent resistor
R3	2.2 Megohm 1/4-watt 5 percent resistor
R4	3.3 Megohm 1/4-watt 5 percent resistor
R5	100K trimpot
R6	500K potentiometer
R7	39K 1/4-watt 5 percent resistor
R8	10K 1/4-watt 5 percent resistor

use a miniature screwdriver-adjust trimpot instead of a front panel potentiometer for this control. The R5 value should be set during calibration, then left alone during actual operation of the circuit.

In noncritical applications, suitable fixed value resistors may be substituted for the two halves of potentiometer R5.

The important feature of this circuit is potentiometer R6. This potentiometer serves as a gain control for the amplifier. Notice that changing the setting of this control alters the current seen by the I-bias input of the OTA. Since the transconductance, or gain of the OTA is set by the I-bias current, the output gain can be manually adjusted via potentiometer R6.

When potentiometer R6 is set for its minimum resistance value, the amplifier's gain will be at its highest. Similarly, adjusting potentiometer R6 for its maximum resistance setting will reduce the amplifier gain to its minimum value.

Using the component values suggested in the parts list, the gain of this amplifier circuit can be adjusted from a low of about 5 to a high of about 100.

Always bear in mind that when high gains are used, the amplitude of the input signal must be limited to avoid signal clipping at the output of the amplifier.

AGC AMPLIFIER

AGC amplifier circuits are used in many advanced radio and television systems. The acronym AGC stands for *automatic gain control*. An AGC amplifier is an amplifier circuit which determines its own gain on the basis of the amplitude of the input signal. The circuit operator does not have to make any manual adjustments.

An AGC amplifier circuit built around one half of an LM13600 chip is illustrated in Fig. 3-6. A suitable parts list for this project is given in Table 3-5.

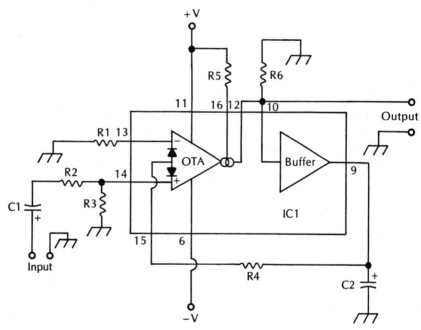

Fig. 3-6 *AGC Amplifier*

**Table 3-5 Parts List for the
AGC Amplifier project of Fig. 3-6.**

Part	Description
IC1	LM13600 dual OTA IC
C1	2µF 35 Volt electrolytic capacitor
C2	50µF 35 Volt electrolytic capacitor
R1, R2	470 ohm 5 percent 1/4-watt resistor
R3	4.7K 5 percent 1/4-watt resistor
R4	180 ohm 5 percent 1/4-watt resistor
R5	22K 5 percent 1/4-watt resistor
R6	33K 5 percent 1/4-watt resistor

This AGC amplifier could also be considered a "compression" amplifier. It effectively compresses, or restricts the range of possible amplitudes. Weak input signals are boosted more than strong input signals.

Using the component values listed in Table 3-5, a 100:1 change in the input amplitude will result in an amplitude change of 5:1 at the circuit's output.

Resistor R4 controls the I-bias current, and thus the overall nominal gain of the amplifier. Since a fixed resistor is being used here, the I-bias value is a constant in this circuit. The value of resistor R6 also affects the circuit's output impedance and signal amplitude.

Notice that only the OTA's noninverting input is being used in this circuit. The inverting input is shorted to ground through resistor R1 for an effective input voltage of zero. Because the noninverting input is being used to accept the input signal, the output signal will be in phase with the input signal.

This circuit is a little unusual compared to most LM13600 circuits. The circuit's output is taken directly from the OTA output, pin 12, instead of from the buffer output, pin 9, but the chip's buffer stage is still being used as an important circuit element. The OTA's output, pin 12, is also fed into the buffer input, pin 10. The buffer's output is then fed back to the linearizing input, pin 15, through resistor R4 and capacitor C2. These two components serve as a simple low-pass filter to help smooth out the buffer's input signal.

In effect, the LM13600's buffer stage in this circuit is functioning as a signal rectifier, producing an I_d current for the OTA's internal linearizing diodes. Normally, there will not be any I_d current to speak of, and the amplifier gain is approximately 40, assuming that the component values listed in Table 3-5 are being used in the circuit. This will hold true unless the OTA's output signal goes high enough because of an increase in the input signal's amplitude to force the Darlington pair buffer to switch on. As the I_d current from the buffer output increases, the OTA's gain is proportionately decreased.

This negative feedback system holds the output amplitude at a more or less continuous level. In other words, the amplifier's gain is being automatically controlled by the instantaneous amplitude of the input signal. Thus, we have an AGC amplifier circuit.

SIMPLE VCA

Another type of amplifier circuit with automated gain is the *Voltage-Controlled Amplifier*, or *VCA*. VCAs are used in many applications, including automation and remote control systems, noise reduction devices, and electronic music synthesizers, among others. Basically, a VCA is an amplifier circuit which varies its gain in proportion to a special control voltage input signal.

A transconductance operational amplifier is an obvious choice for VCA applications, since an OTA is designed to vary its gain, or transconductance, in response to an input current. By feeding the input voltage through a resistor, it will be converted into a current, according to Ohm's Law, and this current can then be fed to the OTA's I-bias input. A simple, but practical VCA circuit built around the CA3080 OTA IC is shown in Fig. 3-7.

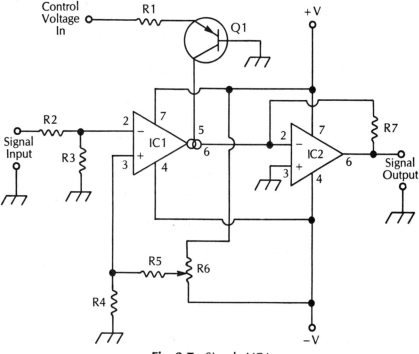

Fig. 3-7 *Simple VCA*

A typical parts list for this project appears as Table 3-6. You may want to breadboard this circuit and experiment with other component values.

48 Amplifier Projects

Table 3-6 Parts List for the Simple VCA project of Fig. 3-7.

Part	Description
IC1	CA3080 OTA IC
IC2	op amp (741, or similar)
Q1	PNP transistor (2N3906, or similar)
R1, R7	10K 5 percent 1/4-watt resistor
R2	100K 5 percent 1/4-watt resistor
R3	1K 5 percent 1/4-watt resistor
R4	100 ohm 5 percent 1/4-watt resistor
R5	1 Megohm 5 percent 1/4-watt resistor
R6	100K trimpot

The control voltage to determine the VCA's gain is applied to the I-bias input, pin 15, through resistor R1 and transistor Q1. The resistor converts the control voltage into a control current, and the transistor functions as a simple linear buffer amplifier stage. The control voltage may be as high as the positive supply voltage used with the circuit.

The input signal is applied to the circuit through a simple voltage divider network made up of resistors R2 and R3. This is done because of the characteristics of the CA3080 IC. This chip gives its best performance if the input signal does not exceed 10 millivolts (0.01 volt). Most practical VCA applications will involve input signals larger than this, so the resistor voltage divider network is used to drop the input signal level down to a value usable by the CA3080. The exact resistor values used here are not particularly crucial. It is only the ratio of the two resistances that matters. By using the resistor values suggested in the parts list (R_2 = 100K and R_3 = 1K), the resistance ratio works out to:

$$100,000/1,000 = 100$$

The input signal level is dropped by a factor of 100. Now the circuit's input signal can range from −1 volt up to +1 volt, without exceeding the CA3080's ±10-millivolt limit.

The CA3080 OTA IC is designed so that its inputs are internally balanced. In most applications, there is no need for an external offset adjustment. But most VCA applications are fairly critical.

Even a small amount of dc offset between the inverting and noninverting inputs can cause problems. For example, in an audio application, such as an electronic music synthesizer, even a little dc offset could result in severe thudding noises when the amplifier is modulated by the external control voltage. This is not just annoying, it could overload and damage loudspeakers or some subsequent audio circuits in the system.

To prevent such problems, some means for dc offset adjustment between the inputs is strongly advisable in a VCA circuit. This is the function of potentiometer R5, and resistor R6. The potentiometer should be a screwdriver-adjust trimpot. Once the circuit has been calibrated, this control should be left alone.

To calibrate this VCA circuit, simply ground the signal input, and adjust trimpot R5 for a true zero output. Changing the control voltage input should not affect the output under these conditions.

For very high precision applications, better calibration of the circuit can be achieved by feeding in a constant, known input signal, and monitoring the output signal on an oscilloscope. Adjust trimpot R5 for a minimum of dc offset in the output signal. That is, the output waveform should be centered around true zero (ground potential).

The output from the transconductance operational amplifier, IC1, is in the form of a current. The buffer amplifier stage built around IC2 serves as a current-to-voltage converter. Almost any standard op amp IC, such as the 741 may be used for IC2 in this circuit. In high-fidelity audio applications, or in any application requiring particularly high precision, it is worth the extra cost to use a high-grade, low-noise op amp here.

Using the component values listed in Table 3-6, the output signal of the entire VCA circuit will have the same amplitude as the original input signal, before being applied to the R2-R3 voltage divider network. That is, if the input signal swings between −1 volt and +1 volt, the output signal will also have a ±1 volt range.

This simple and relatively inexpensive VCA circuit will do the job in many applications. In some cases, however, it may not be quite accurate enough. This is because the control voltage

50 Amplifier Projects

input network, resistor R1 and transistor Q1, is not perfectly linear. The nonlinearity will be most noticeable with small control voltages.

If your intended use demands high precision and high linearity, you will probably be better off with one of the alternative, improved VCA circuits presented in the next few pages.

IMPROVED VCA

An improved VCA circuit is illustrated in Fig. 3-8. Notice that three op amps are used in this circuit. IC1, which is the actual amplifier here, is a CA3080 OTA. IC2 and IC3 are conventional op amp devices. Almost any standard op amp devices may

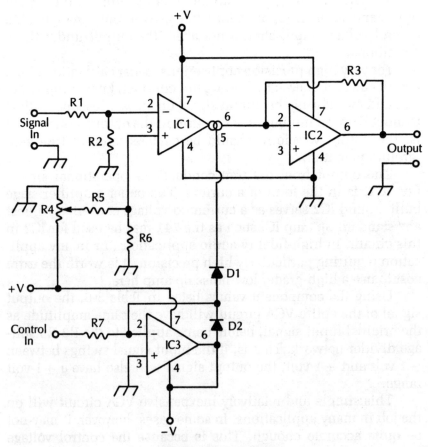

Fig. 3-8 Improved VCA

be employed here. For high precision or high-fidelity audio applications, a high-grade low-noise op amp IC should be used, especially for IC2. Otherwise, virtually any readily available op amp chip will do.

The common 741 IC may be used for IC2 and IC3. You might want to use a dual IC chip for IC2 and IC3 to reduce the project's size and parts count. Some popular dual op amp ICs which can be used in this circuit are the 747, the 1458, and the 4458. Of course, if a dual op amp device is used in the circuit, you will need to correct the pin numbering given in the schematic diagram. A complete suggested parts list for this improved VCA project is given in Table 3-7.

Table 3-7 Parts List for the Improved VCA project of Fig. 3-8.

Part	Description
IC1	CA3080 OTA
IC2	op amp (741, or similar—see text)
D1, D2	diode (1N914, 1N4148, or similar)
R1, R5, R7	22K ¼-watt 5 percent resistor
R2, R6	1K ¼-watt 5 percent resistor
R3	27K ¼-watt 5 percent resistor
R4	100K trimpot

Most of the basic circuitry around IC1 and IC2 is similar to the preceding project. IC2 is a buffer amplifier stage to give the circuit a fairly low output impedance. This stage also converts the OTA's current output into a voltage.

The significant improvement in this version comes from the addition of IC3 and its associated components. This added circuitry makes the control voltage more linear than the simpler approach employed in the circuit of Fig. 3-7. In addition, the control voltage in this circuit is referenced to ground (zero volts). In the simpler VCA circuit presented earlier, the control voltage is referenced to -14.3 volts, because of the internal design of the CA3080's I-bias input.

LM13600 VCA

A third VCA circuit is illustrated in Fig. 3-9. A suitable parts list is given in Table 3-8. This circuit is designed around one half

52 Amplifier Projects

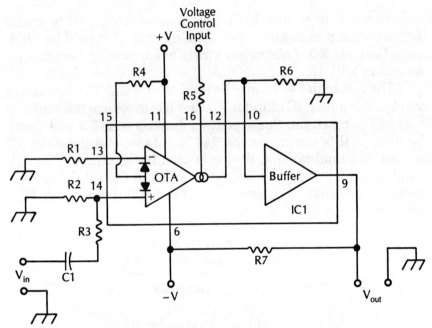

Fig. 3-9 LM13600 VCA

Table 3-8 Parts List for the LM13600 VCA project of Fig. 3-9.

Part	Description
IC1	LM13600 dual OTA IC
C1	0.5 μF capacitor
R1, R2	470 ohm 5 percent 1/4-watt resistor
R3, R6	27K 5 percent 1/4-watt resistor
R4	10K 5 percent 1/4-watt resistor
R5	22K 5 percent 1/4-watt resistor
R7	3.9K 5 percent 1/4-watt resistor

of an LM13600 dual OTA IC. Only a handful of external components are required.

The input signal is fed into the OTA's noninverting input, so the output signal is in phase with the input signal. Capacitor C1 blocks any dc component in the input signal, and resistor R3 limits the input current.

This circuit can easily be modified to use the inverting input instead of the noninverting input, if that suits the intended use.

Simply move capacitor C1 and resistor R3 from pin 14 to pin 13, so the input signal is fed into the inverting input. No other circuit changes are required for this modification.

Resistor R6 loads the OTA's high-impedance output and limits the maximum amplitude of the output signal. The OTA's output signal is then fed into the on-chip buffer stage, resulting in a low impedance output for the circuit as a whole. The buffer's output is loaded by resistor R7.

The control voltage is applied to resistor R5. Thanks to Ohm's Law, the control input is converted into a current that is fed to the IC's I-bias input (pin 16).

The LM13600 IC contains two identical and very closely matched OTAs and buffers. Because the operating characteristics of the two halves of this device are so similar, if not identical, the LM13600 is particularly well suited to stereophonic audio applications. Build two of the circuits shown in Fig. 3-9, and you'll have a stereo (two channel) voltage-controlled amplifier.

In the next few chapters we will be looking at some additional applications for transconductance operational amplifiers that are not primarily intended for signal amplification.

❖4
Switching and Modulation Projects

THE TRANSCONDUCTANCE OPERATIONAL AMPLIFIER PROJECTS IN THIS chapter switch, or modulate, analog signals in a variety of ways.

The combination of switching circuits and modulation circuits might seem a little arbitrary at first glance, but electronic switching is actually a form of crude modulation. The modulated signal is turned completely on or completely off.

Any modulation circuit has two inputs. One input is known as the *program signal*. This signal contains the information to be transmitted or stored. The other input signal is the *carrier signal*, which defines just how the program signal will be modulated.

There are many different possible approaches to modulation. Almost any parameter of a program signal may be modulated to suit the specific requirements of the intended use. As you will learn in this chapter, an OTA is well suited to a great many different types of modulation.

ANALOG SWITCH

A practical analog switch circuit built around a CA3080 OTA IC is illustrated in Fig. 4-1. A suitable parts list for this project appears in Table 4-1.

In many respects, this circuit resembles a modified VCA. Instead of a continuous range of controlling voltages, and resulting amplifier gains, this circuit is designed to accept control signals in the form of pulses to *gate* the amplifier ON and OFF. Either the amplifier gain is at its maximum, gate open—switch on, or the gain is at its minimum, gate closed—switch off.

56 Switching and Modulation Projects

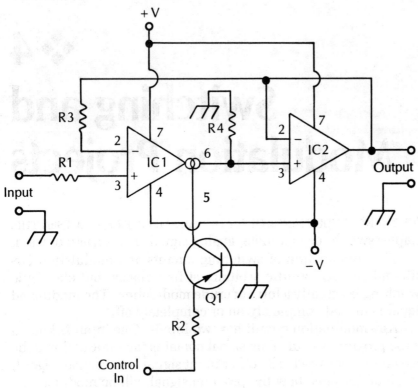

Fig. 4-1 Analog Switch

**Table 4-1 Parts List for the
Analog Switch project of Fig. 4-1.**

Part	Description
IC1	CA3080 OTA
IC2	op amp (TL071, or similar—see text)
Q1	PNP transistor (2N2907, or similar)
R1, R3	10K ¼-watt 5 percent resistor
R2	18K ¼-watt 5 percent resistor
R4	47K ¼-watt 5 percent resistor

When the switch circuit is in its ON state, the input signal can reach the output. The amplifier is set up for unity gain. When the switch circuit is in its OFF state, the input signal is blocked from the output. As you can see, this circuit very closely simulates the action of a mechanical SPST switch, except it is under electronic, rather than mechanical control. The switch is turned ON and OFF with a voltage pulse, instead of by the mechanical motion of a slider.

IC1, the OTA, is a simple noninverting amplifier. It is followed by IC2, which is configured as a noninverting buffer amplifier. This stage is included in the circuit to convert the OTA's output current into an output voltage, and to reduce the circuit's output impedance.

Transistor Q1 and resistor R2 make up a simple voltage-controlled current switch. When the transistor is ON, current can flow through its collector, and thus into the I-bias pin (5) of the OTA (IC1). When transistor Q1 is cut OFF, however, no current can flow through the collector, so the value of I-bias is effectively zero, so the OTA is gated OFF. The circuit's main input signal is blocked from the output.

You may well want to experiment with alternative values for resistor R2. This resistance determines the required ON voltage. Simply determine what input control voltage you want to use to turn the switch circuit ON, and multiply that voltage by approximately 2000 ohms. The resulting value will be the required resistance for resistor R2.

In the parts list of Table 4-1, a value of 18K is given for resistor R2. This means the control voltage needed to turn the circuit on is:

$$V_c = R_2/2000$$
$$= 18000/2000$$
$$= 9 \text{ volts}$$

This control voltage will be suitable for many applications, since 9 volts is a commonly used voltage in both analog and CMOS digital circuits.

For other practical applications, you may be working with other switching voltages, so you will need to change the value of resistor R2 accordingly. For example, if you want to control this analog switch circuit from a TTL digital circuit, the switching voltage will need to be 5 volts, so resistor R2's value should be changed to:

$$R_2 = V_c \times 2000$$
$$= 5 \times 2000$$
$$= 10,000 \text{ ohms}$$
$$= 10K$$

Almost any electronic signal, either analog or digital, may be switched with this circuit, provided that the maximum limits of the OTA are not exceeded. Refer to chapter 2, or the manufacturer's specification sheet.

This analog switch circuit offers several advantages in certain applications over CMOS analog switch ICs such as the CD4066 and CD4016. These chips require the switching voltage to be close to the circuit's supply voltage. This project does not suffer from that limitation. Also, unlike these CMOS devices, this circuit, is not limited to a 15-volt signal range.

FAST-INVERTING SWITCH

Another analog switch circuit built around the CA3080 OTA IC is illustrated in Fig. 4-2. As you can see, this is quite a simple

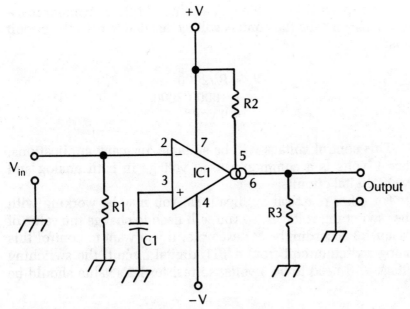

Fig. 4-2 Fast-inverting Switch

circuit. Just four external components are required in addition to the IC itself.

A suitable parts list for this project is given in Table 4-2. The OTA's inverting input is used to accept the external input signal in this circuit.

Table 4-2 Parts List for the Fast-Inverting Switch project of Fig. 4-2.

Part	Description
IC1	CA3080 OTA
C1	0.01 µF capacitor
R1, R3	10K ¼-watt 5 percent resistor
R2	2.2K ¼-watt 5 percent resistor

Notice that there is no control input in this switching circuit. The switching function is self-automated here. If the instantaneous input voltage at the inverting input is less than the reference voltage at the noninverting input, the circuit's output will be LOW. If, on the other hand, the instantaneous input voltage at the inverting input exceeds the reference voltage at the noninverting input even slightly, the circuit's output will go HIGH.

This circuit is actually a form of *comparator*. Additional comparator circuits will be presented later in this chapter. The OTA's noninverting input is referenced to ground through capacitor C1.

The value of resistor R2 is selected to give a fairly high I-bias current—in the range of several hundred µA. With a transconductance in this range, the CA3080's slew rate is approximately 20 volts per microsecond. This means the switching circuit can respond very quickly to changes in the input signal.

If the input voltage at the OTA's inverting input is very close to the reference voltage at the noninverting input, the OTA will function as a high gain linear amplifier. When the input voltage is significantly different from the reference voltage, the circuit's output will be limited by the I-bias value, and thus, by the value of resistor R2.

Try experimenting with alternative values for this resistor. It is best to keep this resistance within the 1K to 10K range. If resistor R2 has a value of 1K, the circuit's output voltage will be limited to about 0.7 volt. At the other extreme, using a 10K resistor for R2 will allow the output voltage to extend as high as 7 volts. Using the 2.2K resistor called for in the parts list will set the output voltage limit at a point a little over 1 volt.

SCHMITT TRIGGER

Schmitt triggers are specialized types of switching circuits which are frequently used to "clean up" noisy signals after transmission, or for storage and retrieval. Figure 4-3 illustrates how the CA3080 can be put to work as a Schmitt trigger. A suitable parts list for this project is given in Table 4-3. Notice that unequal supply voltages are used in this circuit.

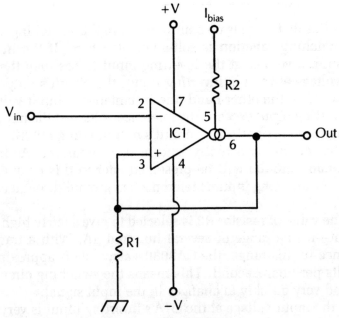

Fig. 4-3 Schmitt Trigger

Table 4-3 Parts List for the Schmitt Trigger project of Fig. 4-3.

Part	Description
IC1	CA3080 OTA
R1, R2	10K 1/4-watt 5 percent resistor

Figure 4-4 illustrates how a Schmitt trigger cleans up a noisy signal. The Schmitt trigger's output is either HIGH or LOW, like a digital circuit. The circuit is designed to recognize a specific switchover voltage. If the input signal is less than the circuit's switchover voltage, then the output will be LOW. If the input signal voltage is greater than the switchover voltage, the Schmitt trigger's output goes HIGH. The bulk of the noise and garbage in the input signal is effectively ignored.

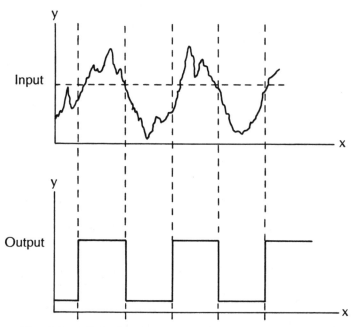

Fig. 4-4 *A Schmitt trigger can "clean up" a noisy signal.*

A Schmitt trigger can also be used to convert any analog waveshape into a pulse form usable by digital circuitry, as illustrated in Fig. 4-5. This is one of the simplest possible circuits using a transconductance operational amplifier. Aside from the OTA chip itself, the only external components required in this circuit are a pair of resistors.

In very basic terms, this Schmitt trigger circuit is a modified comparator. (Comparator circuits will be discussed later in this chapter). The input voltage, which is applied to the OTA's inverting input is compared to a reference voltage at the noninverting input. In this circuit, the noninverting input is grounded through resistor R1.

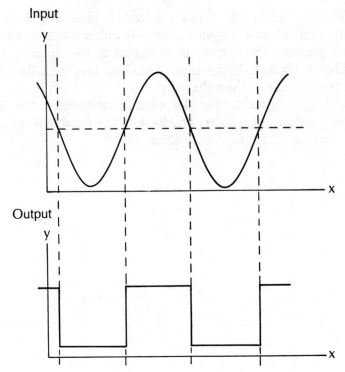

Fig. 4-5 *A Schmitt trigger can convert an analog signal into a digital pulse form.*

Because the inverting input is being used to accept the input signal voltage, the output polarity is inverted by this circuit.

If the instantaneous input voltage is less than the reference voltage, the OTA's output will be HIGH, with a positive output voltage equal to the product of the values of the I-bias current and resistor R2.

If, at any time, the input voltage exceeds the reference voltage, the OTA's output will very rapidly switch into a LOW state. Now the output voltage will be negative, again with a value that is equal to the product of the values of the I-bias current and resistor R2.

This Schmitt trigger circuit is somewhat sensitive to noise in the input signal. If the input voltage is very close to the reference voltage, a brief burst of noise in the signal could cause the circuit to falsely trigger. This can result in a condition known as "output chatter." The output may oscillate rapidly back and forth between the HIGH and LOW states under these conditions.

In effect, the circuit gets "confused" by the noise. If this is a problem in your particular application, you will need a Schmitt trigger circuit with greater hysteresis. A project of this type will be presented a little later in this chapter.

LOW-POWER SCHMITT TRIGGER

An alternative Schmitt trigger circuit is illustrated in Fig. 4-6. A suitable parts list for the this project appears as Table 4-4.

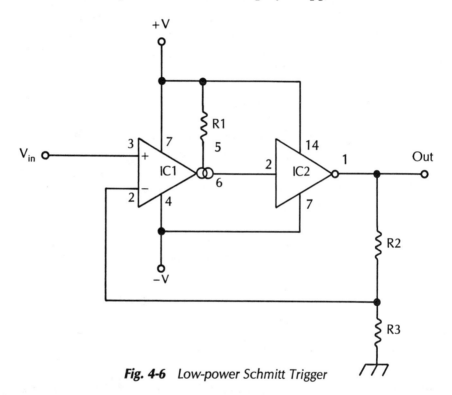

Fig. 4-6 Low-power Schmitt Trigger

Table 4-4 Parts List for the Low-power Schmitt Trigger project of Fig. 4-6.

Part	Description
IC1	CA3080 OTA
IC2	CD4049 CMOS inverter
R1	100K 1/4-watt 5 percent resistor
R2	22K 1/4-watt 5 percent resistor
R3	10K 1/4-watt 5 percent resistor

This circuit offers several advantages over the simple Schmitt trigger circuit of Fig. 4-3. This circuit consumes less power, and the output is suitable for use with digital CMOS circuitry. The output voltage switches between the circuit's positive and negative supply voltages. Moreover, the switching threshold voltages are more flexible in this circuit. They can be changed by altering simple resistor values.

For proper operation, the supply voltages for this circuit should not be permitted to exceed ±9 volts. The switching threshold point for this Schmitt trigger circuit is determined by the values of resistors R2 and R3, and the circuit's positive supply voltage. The formula for determining the switching threshold voltage is as follows:

$$V_s = V+ \times R_3 / (R_2 + R_3)$$

When the input voltage crosses through the V_s voltage, the circuit's output reverses states. Assuming a ±9-volt power supply, and using the component values suggested in the parts list for this project (Table 4-4), the switching threshold voltage (V_s is approximately equal to:

$$R_2 = 22000 \text{ ohms}$$
$$R_3 = 10000 \text{ ohms}$$
$$\begin{aligned} V_s &= V+ \times R_3 / (R_2 + R_3) \\ &= 9 \times 10000 / (22000 + 10000) \\ &= 9 \times 10000 / 32000 \\ &= 9 \times 0.3125 \\ &= 2.8 \text{ volts} \end{aligned}$$

By all means, you are encouraged to experiment with alternative values for resistors R2 and R3.

VARIABLE HYSTERESIS SCHMITT TRIGGER

An important specification for any Schmitt trigger circuit is the *hysteresis*. In very simple terms, hysteresis is a sort of delay in the circuit's response. It might be considered as a form of sluggishness. Common sense would seem to indicate that the less hysteresis, the better; but in many practical applications, the conclusions of common sense don't hold up.

If a noisy input signal is close to the switchover value, the Schmitt trigger will be prone to "chatter", as shown in Fig. 4-7. By increasing the hysteresis of the circuit, the turn-ON switchover voltage will be higher than the switch-OFF voltage. This leaves a "dead zone" between the two switching levels, which is ignored by the circuit. Minor noise pulses near the switchover voltages have considerably less effect under these circumstances.

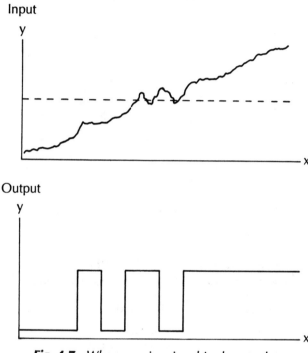

Fig. 4-7 *When a noisy signal is close to the switchover value, output chatter may result.*

There is a definite tradeoff involved with hysteresis, however. A low-hysteresis circuit will respond faster to smaller changes in the input signal, while a high-hysteresis circuit is less sensitive to noise. The ideal amount of hysteresis will depend on the requirements of the specific intended use.

The circuit shown in Fig. 4-8 features a variable amount of hysteresis. A suitable parts list for this project is given in Table 4-5.

Adjusting potentiometer R5 controls the I-bias value being fed into pin 5 of the OTA (IC1). The greater the I-bias value, the

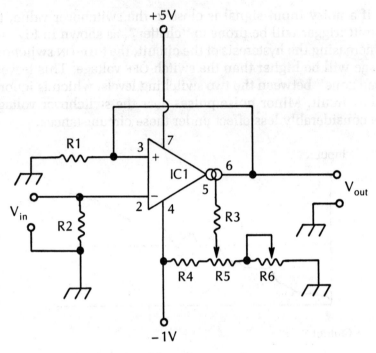

Fig. 4-8 Variable Hysteresis Schmitt Trigger

**Table 4-5 Parts List for the
Variable Hysteresis Schmitt Trigger project of Fig. 4-8.**

Part	Description
IC1	CA3080 OTA IC
R1	47K 5 percent 1/4-watt resistor
R2	100K 5 percent 1/4-watt resistor
R3	33K 5 percent 1/4-watt resistor
R4, R5	12K 5 percent 1/4-watt resistor
R6	25K potentiometer

greater the hysteresis exhibited by the circuit. Of course, reducing the I-bias value also reduces the hysteresis.

In some applications, it may be appropriate to use the potentiometer as a manual front panel control. For most uses, however, a screwdriver adjust trimpot might be more suitable. Calibrate the circuit to the desired amount of hysteresis, and then leave the control alone.

SAMPLE-AND-HOLD CIRCUIT

A *Sample-and-Hold*, or S-H circuit is a specialized form of automated analog switch. The input signal is almost always in analog form. There would be little point in inputting a digital signal to an S-H circuit. The switch is activated at regular intervals by a pulse-wave oscillator. Each time the switch is activated, the instantaneous voltage value of the input signal is measured. This instantaneous voltage is stored in a capacitor and held until the next control pulse activates the switch again.

The operation of a typical Sample-and-Hold circuit is illustrated in Fig. 4-9. A practical Sample-and-Hold circuit built

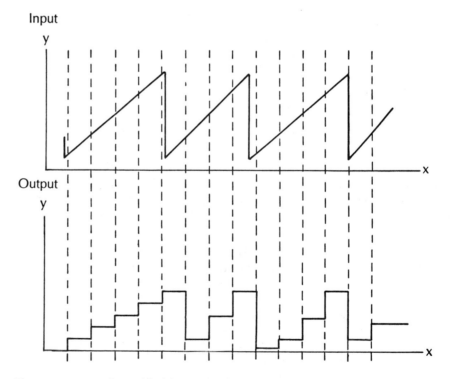

Fig. 4-9 A sample-and-hold circuit takes periodic instantaneous "readings" of an input signal and holds each value until the next sample is taken.

around the CA3080 OTA IC is shown in Fig. 4-10. Notice that this circuit is basically a more sophisticated and slightly more complex version of the analog switch circuit presented in Fig. 4-1.

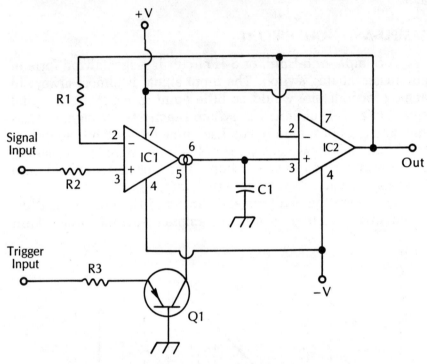

Fig. 4-10 *Sample-and-Hold Circuit*

A suitable parts list for this Sample-and-Hold project appears in Table 4-6. IC1 is the OTA (CA3080), which serves as the actual switching amplifier. IC2, which may be almost any conventional op amp, serves as a buffer amplifier. The buffer is

Table 4-6 Parts List for the Sample-and-Hold circuit of Fig. 4-10.

Part	Description
IC1	CA3080 OTA
IC2	op amp (CA3140, or similar)
Q1	PNP transistor (2N2907, or similar)
C1	0.5 µF capacitor
R1, R2	10K ¼-watt 5 percent resistor
R3	22K ¼-watt 5 percent resistor

particularly vital in this circuit to prevent loading and premature discharge of the holding capacitor (C1).

For best results, a high-grade op amp with a high impedance FET input is strongly recommended for IC2. A common 741 could be used, but the circuit's performance will suffer some, especially in audio applications. The CA3140 op amp IC is a good choice for the buffer amplifier in this circuit.

Transistor Q1 and resistor R3 condition the control voltage input and convert it into a suitable I-bias current. The characteristics of the transistor are not terribly crucial, and it should be OK to substitute another low-power PNP type transistor for Q1.

The OTA's noninverting input is used for the signal input, so the output polarity is not inverted, although this isn't of too much significance in most Sample-and-Hold applications.

Probably the most important component in this circuit is the holding capacitor, C1. This capacitor stores the held sample, and continues feeding this value to the circuit output through buffer amplifier IC2 until the next sample is taken. The capacitance of C1 should not be too large, or the capacitor may not charge up to the full sample value fast enough. On the other hand, if the capacitance is too small, the capacitor won't be able to hold the full sampled voltage for very long. I feel $0.5\mu F$ is a good compromise value for most practical Sample-and-Hold applications.

Use a high-grade, low-leakage capacitor for C1. Polyester, mylar, or polystyrene capacitors would be good choices for use in this circuit. If you need sampled values to be held for relatively long periods, you may increase the capacitance of C1. If you use a capacitor with more than about 1 μF, you should use a tantalum device. Electrolytic capacitors will not perform satisfactorily in this type of circuit.

An internal clock source or control pulse generator is not included in the circuit of Fig. 4-10. Almost any rectangular-wave generator may be used as an external control pulse input source.

One of the simplest control pulse sources is shown in Fig. 4-11. This is just a simple narrow-width rectangular-wave generator circuit built around an inexpensive 555 timer IC. A suitable parts list for this oscillator circuit is given in Table 4-7. Using the component values given here, the input signal will be sampled about once every other second (0.5 Hz). You may want to experiment with other component values to suit various other applications.

70 Switching and Modulation Projects

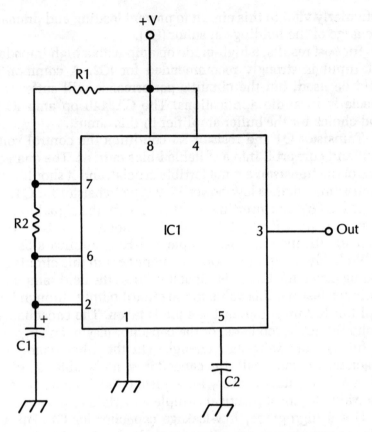

Fig. 4-11 This circuit can be used to generate the trigger pulses for the Sample-and-Hold circuit of Fig. 4-10.

Table 4-7 Parts List for the Pulse Generator circuit of Fig. 4-11.

Part	Description
IC1	555 timer IC
C1	50 µF 35 Volt electrolytic capacitor
C2	0.01 µF capacitor
R1	1.2K 5 percent 1/4-watt resistor
R2	27K 5 percent 1/4-watt resistor

Sample-and-Hold circuits are often used in electronic music systems. They are also employed in some signal transmission and data recording systems, and analog-to-digital (A-D) converters.

INVERTING-VOLTAGE COMPARATOR

A *comparator* is a switching circuit which compares two signals, usually in the form of voltages, and produces an output that indicates which of the two input signals is the larger. There are obviously three possible combinations of two inputs:

$$A > B$$
$$A = B$$
$$A < B$$

Any combination of inputs must fit into one of these three categories.

One of the two input signals to a comparator is usually called the *reference voltage*, or V_{ref}. This voltage is usually held at a constant value, although this is not true in all applications.

The other input is considered the signal input, and is normally labeled V_{in}. This signal usually varies over time. The input voltage (V_{in}) is continuously compared with the reference voltage (V_{ref}), and the comparator circuit's output always indicates the present relation of the two inputs.

A practical comparator circuit built around the CA3080 OTA IC appears in Fig. 4-12, with a suitable parts list given in Table 4-8.

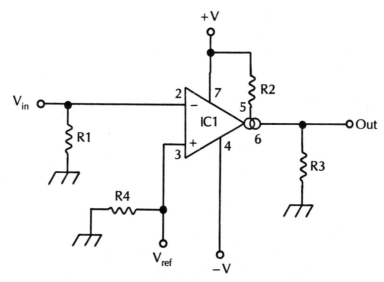

Fig. 4-12 Inverting-voltage Comparator

Table 4-8 Parts List for the
Inverting-Voltage Comparator project of Fig. 4-12.

Part	Description
IC1	CA3080 OTA
R1, R3, R4	10K 1/4-watt 5 percent resistor
R2	2.2K 1/4-watt 5 percent resistor

The varying input signal, V_{in}, is applied to the OTA's inverting input, and the reference voltage is fed to the noninverting input.

This circuit is designed to exhibit a very high gain. Resistor R2 is selected to give a rather large I-bias current, resulting in a high gain in the OTA. The I-bias value in this circuit is on the order of several hundred microamps. This combination of the I-bias and R2 values results in a slew rate of approximately 20 volts per microsecond, so the circuit has a very rapid response.

The gain is high enough that the linear region of the output response is very narrow. Unless the two input voltages are quite close in value without being exactly equal, the high gain will cause the OTA to function as a true comparator. If the input voltage (V_{in}) is significantly different from the reference voltage (V_{ref}), the output will be at either the maximum positive value or the maximum negative value.

The value of resistor R2 will determine the maximum limits of the output voltage. Do not give resistor R2 a value larger than 10K. Using the 2.2K resistor specified in the sample parts list, the output will be limited to a value of approximately ±1.5 volts.

The operation of this comparator circuit can be summarized as follows:

Input Combination	Output
$V_{in} > V_{ref}$	−1.5 volts
$V_{in} = V_{ref}$	0 volt
$V_{in} < V_{ref}$	+1.5 volts

Notice that since the input signal, V_{in}, is applied to the OTA's inverting input, the output polarity is reversed.

If V_{in} is close to, but not exactly equal to V_{ref}, the output will be a nonzero voltage, but not quite at the ±1.5-volt limit. This will be a very narrow range, thanks to the high gain of the amplifier in this circuit, and it should not be a problem in most applications. The output of this circuit is graphed in Fig. 4-13.

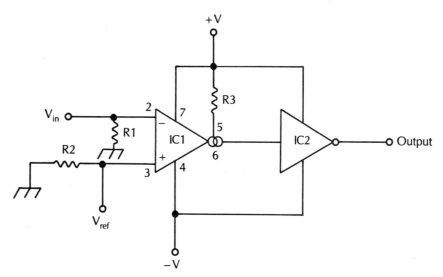

Fig. 4-13 This graph illustrates the operation of the circuit shown in Fig. 4-12.

NONINVERTING VOLTAGE COMPARATOR

In some applications, the polarity inversion of the comparator circuit of Fig. 4-12 will be undesirable. In such cases, we can use a noninverting voltage comparator circuit, like the one illustrated in Fig. 4-14. A typical parts list for this project appears in Table 4-9.

This comparator circuit is extremely sensitive, with a very narrow questionable range. Another nice feature of this circuit is that the power consumption is quite low, making the circuit well suited to large systems, and portable (battery powered) applications.

Notice that the input signal (V_{in}) is still being applied to the OTA's inverting input, but the OTA's output signal is then re-inverted by IC2, a CMOS digital inverter.

The output voltage of this circuit swings from the positive supply voltage to the negative supply voltage. The differences between the output states are much more obvious. This circuit can also be used to directly drive CMOS digital circuits.

74 Switching and Modulation Projects

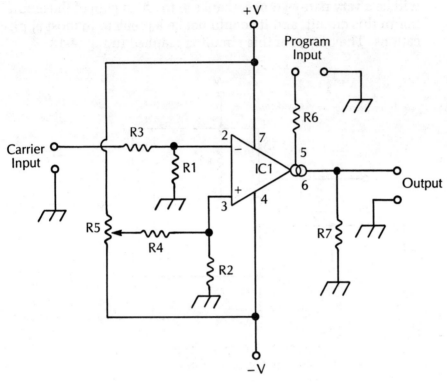

Fig. 4-14 Noninverting Voltage Comparator

Table 4-9 Parts List for the Noninverting Voltage Comparator project of Fig. 4-14.

Part	Description
IC1	CA3080 OTA
IC2	CD4049 CMOS inverter
R1, R2	10K 1/4-watt 5 percent resistor
R3	1 Megohm 1/4-watt 5 percent resistor

In this circuit, the I-bias current is fixed at a little under 20 µA. The OTA's output is fed into the CMOS inverter, which exhibits a nearly infinite input impedance. Under these conditions, the amplifier gain is greater than 125 dB. This means that if the two input signals (V_{in}) and V_{ref}) are different by only a few microvolts, the circuit's output will be at one of its two extremes (positive or negative). The in-between range is so narrow it is virtually negligible for all but the most extremely critical applications.

This circuit typically consumes less than 50 μA, so it will not present much of a drain to most practical power supplies.

CA3080 AMPLITUDE MODULATOR

As stated earlier in this chapter, switching circuits represent a crude and simple form of modulation. For the remainder of this chapter we will be working with true modulation circuits of a more sophisticated type.

A true modulation system uses two input signals. One is the *carrier signal*. This is usually a more or less constant waveform, and is generally of a relatively high frequency. In radio transmissions, the carrier signal is the RF frequency the radio is tuned to. In analog modulation systems, the carrier signal is usually a sine wave because of the simplicity of this particular waveform. Unlike all other waveforms, a sine wave consists of a single-frequency component.

In digital and pseudo-digital modulation systems, a pulse wave of some sort is used for the carrier signal.

The second input signal in a modulation system is the *program signal*. This signal carries the actual intelligence or information to be transmitted or stored and later retrieved. It might be music, speech, or electrically encoded data.

The program signal generally has a much lower frequency than the carrier signal. The program signal is superimposed over the carrier wave, modulating some parameter of the waveform. Almost any parameter of the carrier wave may be modulated. Typical modulation systems use *amplitude modulation (AM)*, *frequency modulation (FM)*, and *phase modulation (PM)*. Projects for all of these types of modulation will be presented in the following pages.

An *amplitude modulator* superimposes the amplitude of the program signal over the carrier wave. The level of the modulation carrier signal varies over time, in step with the program.

Amplitude modulator circuits are also often known as *two-quadrant multipliers*. In effect, the program signal is multiplied by the carrier signal.

76 Switching and Modulation Projects

In a sense, an amplitude modulator is very similar to a VCA. (See chapter 3.) The control voltage (program) determines the instantaneous amplitude of the input signal (carrier). The primary difference between an amplitude modulator and a VCA is in how the circuits are used.

The CA3080 OTA IC is used as the heart of the amplitude modulator circuit shown in Fig. 4-15. A suitable parts list for this project is given in Table 4-10.

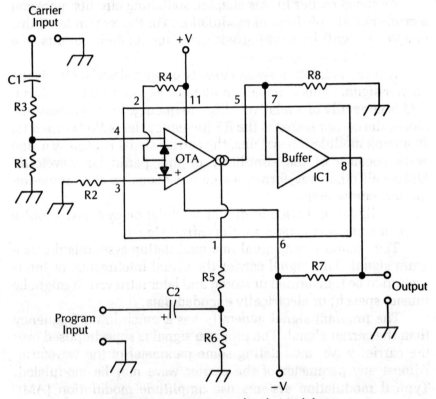

Fig. 4-15 *CA3080 Amplitude Modulator*

As you can see, this circuit is not particularly complex. The only external components required for this project are a handful of simple resistors.

The carrier input is fed into the OTA's inverting input. The noninverting input is grounded through resistor R2. Potentiometer R5 is used to adjust for offset null, that is, a true zero output, when the input is zero. This control should be a screwdriver

**Table 4-10 Parts List for the
CA3080 Amplitude Modulator project of Fig. 4-15.**

Part	Description
IC1	CA3080 OTA
R1, R2	100 ohm 1/4-watt 5 percent resistor
R3	56K 1/4-watt 5 percent resistor
R4	47K 1/4-watt 5 percent resistor
R5	100K trimpot
R6	33K 1/4-watt 5 percent resistor
R7	100K 1/4-watt 5 percent resistor

adjust trimmer potentiometer. It should be set during calibration, then left alone during normal use of the circuit.

Because the inverting input is being employed in this circuit, the carrier signal is phase shifted 180 degrees at the output. This phase shift is completely irrelevant in virtually all amplitude modulation applications.

The program signal, which is to modulate the carrier signal, is fed into the I-bias input (pin 5) through resistor R6. The amplifier gain varies from instant to instant, depending on the instantaneous amplitude of the program signal. As a result, the amplified carrier wave at the output continuously changes in level, in step with the program signal.

The overall nominal gain of the circuit is most directly controlled by the value of resistor R3. Try experimenting with alternative values for this component. A resistance of about 35K will result in approximately unity gain, depending on the other component values used in the circuit.

Resistors R1 and R2 serve to bias the OTA's inputs. Very low values are used for these resistors, so there is no need for an external slew rate limiting capacitor in this circuit. The low-value input bias resistors also help reduce the noise levels of the IC.

The program signal input can range from the positive supply voltage to the negative supply voltage. A power supply of ±9 volts is recommended for this project. When the modulation (program) input is at its maximum level (V+), the amplifier's gain is about double its nominal (zero volt modulation) value. Dropping the program signal to its minimum level (V−) cuts the carrier wave's amplitude about 80 dB from its original input level. As you can see, the carrier signal can be modulated over a very wide range.

LM13600 AMPLITUDE MODULATOR

An alternative amplitude modulator circuit is illustrated in Fig. 4-16. This project is built around one-half of an LM13600 dual OTA IC. Both halves of the chip can be used in stereo applications. Be sure to double-check the pin numbers. A suitable parts list for this project appears in Table 4-11.

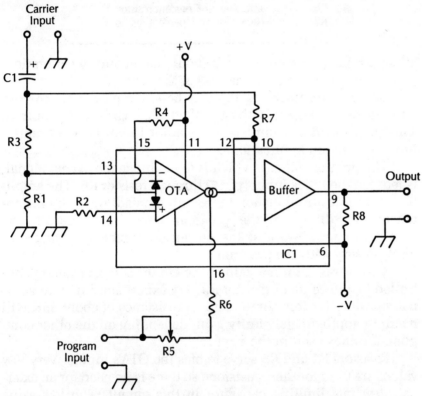

Fig. 4-16 *LM13600 Amplitude Modulator*

As in the preceeding circuit, the carrier input is fed into the OTA's inverting input, with the noninverting input grounded. No offset null control is required for this particular circuit.

The program (modulation) input signal is fed into the I-bias input (pin 16) through capacitor C2 and resistor R5. The capacitor blocks any dc component of the program signal, and the resistor converts the input voltage into an input current.

Table 4-11 Parts List for the LM13600
Amplitude Modulator project of Fig. 4-16.

Part	Description
IC1	LM13600 dual OTA IC
C1	0.5 µF capacitor
C2	2 µF 35 Volt electrolytic capacitor
R1, R2	470 ohm 5 percent ¼-watt resistor
R3, R8	27K 5 percent ¼-watt resistor
R4, R6	10K 5 percent ¼-watt resistor
R5	22K 5 percent ¼-watt resistor
R7	4.7K 5 percent ¼-watt resistor

If the program input is at its lowest value, approximately V−, the amplifier gain will be very close to zero. To all intents and purposes, the carrier input signal will be blocked from the circuit's output. As the program input voltage becomes increasingly positive, the amplifier's gain is increased proportionately.

In some modulation applications, it may be desirable to dc bias the program signal, so that the gain is not zero when the program signal is at zero. It is very easy to add a dc bias to this circuit. Just feed the appropriate bias voltage into the junction of resistors R5 and R6, past the input capacitor (C1).

The polarity of the program input signal controls the phase shift of the carrier signal. If the program voltage is positive, the carrier wave at the circuit's output will be phase shifted 180 degrees, or inverted in comparison with the original carrier input signal. On the other hand, a negative program input voltage will result in the carrier wave at the output being in phase with the original carrier signal at the input.

Amplitude modulation generates *sidebands*, or frequency components which aren't part of either of the original input signals (carrier and program). These sidebands are at frequencies equal to the sums and differences of each pair of frequency components in the carrier and program waveforms. The sidebands appear on higher and lower sides of the carrier wave at the output.

The output waveform from an amplitude modulation circuit is always more complex than either of the input signals (carrier and program) being fed into the circuit.

AM SIDEBANDS

Before moving on to the next project, let's take a few minutes for a closer look at the sidebands produced by the amplitude modulation process.

Whenever amplitude modulation is used, new, "phantom" frequency components are added to the output signal. These added frequency components are not part of either of the original carrier or program input signals. Generated by the modulation process itself, these extra frequency components are known as *sidebands*. Their relative strength, or amplitude with respect to the carrier wave at the output, is determined by the amount of modulation involved.

The sidebands can not occur at just any frequency. The sideband frequencies are always equal to the sum(s) and difference(s) of all of the frequency components in the carrier and program input signals.

To keep things simple, we will assume that both the carrier signal and the program signal are sine waves in the following example. A sine wave is the simplest possible ac waveform. It consists of just a single frequency component, known as the *fundamental*.

Let's assume that the carrier frequency is 12 kHz, and the program signal is a continuous sine wave with a frequency of 5.5 kHz. If this program signal is used to modulate the carrier signal, two sidebands will be generated:

$$C + P = 12{,}000 + 5500$$
$$= 17{,}500 \text{ Hz}$$

and:

$$C - P = 12{,}000 - 5500$$
$$= 6500 \text{ Hz}$$

Except for sine waves, all other repeating waveforms include additional frequency components known as *harmonics*. A harmonic is a frequency component that is a whole number multiple of the fundamental frequency. For example, let's assume that the fundamental frequency is 250 Hz. The harmonic series for this fundamental looks like this:

Fundamental	250 Hz	
Second harmonic	500 Hz	(× 2)
Third harmonic	750 Hz	(× 3)
Fourth harmonic	1000 Hz	(× 4)
Fifth harmonic	1250 Hz	(× 5)
Sixth harmonic	1500 Hz	(× 6)
Seventh harmonic	1750 Hz	(× 7)
Eighth harmonic	2000 Hz	(× 8)
Ninth harmonic	2250 Hz	(× 9)
Tenth harmonic	2500 Hz	(× 10)

and so forth.

Not all waveforms include all of the harmonics. A triangular wave, for example, includes only the fundamental and the odd harmonics:

Fundamental
Third harmonic
Fifth harmonic
Seventh harmonic
Ninth harmonic

and so on.

Another difference between waveforms is the relative amplitude of the harmonics, compared to the fundamental. For example, a square wave has the same series of harmonics as a triangular wave, but the harmonic content is much stronger in the square wave.

When complex waveforms (those including harmonics) are used in amplitude modulation, each frequency component in the program signal generates its own pair of sidebands for every frequency component in the carrier signal. This is why the carrier wave is usually a sine wave. If a complex multiple frequency component waveform is used for the carrier, the modulated output will have a very large and confusing number of sidebands. The signal will be harder to demodulate, and each generated sideband necessarily uses up a finite amount of energy. A complex, non-sine carrier wave is almost always wasteful, at best.

The one exception to this general rule of always using a sine wave for the carrier is in electronic music or sound effect systems. Using a non-sine carrier wave can result in some very complex sounds, which might be useful in certain cases.

RING MODULATOR

Assuming that only sine waves are used as input signals, the output signal of an amplitude modulator is comprised of four frequency components:

- Program
- Carrier
- Carrier − Program
- Carrier + Carrier

A variation on the basic amplitude modulator is the *ring modulator*. This type of circuit works similarly to an amplitude modulator, except the original input signals are suppressed at the output. The output consists of just the sidebands generated by the modulation process.

Again, if we assume that only sine waves are being used as the input signals, the output signal of a ring modulator will be made up of just two frequency components:

- Carrier − Program
- Carrier + Program

An amplitude modulator circuit is sometimes called a *two-quadrant multiplier*. Similarly, a ring modulator is occasionally referred to as a *four-quadrant multiplier*.

A practical ring modulator circuit built around one-half of an LM13600 dual OTA IC is shown in Fig. 4-17. A suitable parts list for this project is given in Table 4-12.

Notice how similar this circuit is to the amplitude modulator circuit of Fig. 4-16. The biggest differences are in the component values used and the feedback path, through resistor R7, from the OTA's output (pin 12) to its inverting input (resistor R3 and pin 13).

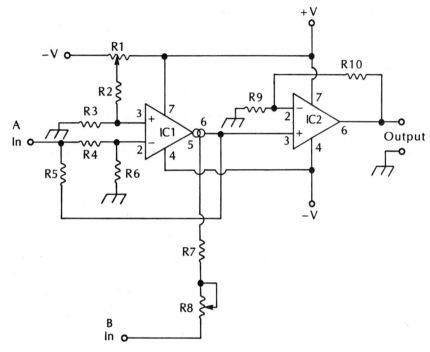

Fig. 4-17 Ring Modulator

Table 4-12 Parts List for the Ring Modulator project of Fig. 4-17.

Part	Description
IC1	LM13600 dual OTA IC
C1	2 µF 35 Volt electrolytic capacitor
R1, R2	470 ohm 5 percent ¼-watt resistor
R3	4.7K 5 percent ¼-watt resistor
R4, R6, R7, R8	10K 5 percent ¼-watt resistor
R5	50K potentiometer

If the program input is grounded (modulation voltage of zero volts), the carrier signal is effectively blocked from the output. A program signal of zero results in a circuit output of zero.

The amplifier's gain increases as the program voltage moves away from zero in either direction, positively or negatively. The polarity of the modulation signal determines the phase of the output. If the program input is negative, the output signal will be in phase with the carrier input signal. On the other hand, if the program input is positive, the output signal will be inverted. That is, the output will be 180 degrees out of phase with the carrier input signal.

Potentiometer R5 permits the circuit operator to set the circuit's sensitivity to the program signal. That is, this potentiometer determines how large the program voltage must be for a given amount of modulation. In some applications, this potentiometer may be replaced with a fixed resistor.

Potentiometer R5 may also be considered a calibration control. The potentiometer setting is adjusted for a true zero circuit output when the program input is grounded.

The carrier signal is fed into the OTA's inverting input, so naturally, the OTA's output at pin 12 is inverted. With a zero volt program input, the OTA output should be exactly equal to the carrier input at the inverting input (pin 13). The OTA's output is fed back through resistor R7 to the inverting input, and the two signals cancel each other, leaving a zero output.

When the program input voltage is positive, the OTA's gain is increased, so its output signal is larger than the original carrier input signal. As a result, an inverted signal is generated. The amplitude of this signal will depend on the exact amount of gain, which is, in turn, dependent on just how positive the program input voltage happens to be.

On the other hand, if the program input voltage is negative, the OTA's gain is decreased. The feedback signal is no longer enough to cancel the original carrier input signal, and a noninverted signal appears at the circuit's output.

This particular circuit works best with a fairly large set of power supply voltages. It is recommended that you operate this circuit off of a ±15-volt power supply.

FOUR-QUADRANT MULTIPLIER

A somewhat different four-quadrant multiplier circuit is shown in Fig. 4-18. This circuit is built around the CA3080 OTA IC (IC1). A conventional op amp, such as the 741 is used as an output buffer stage (IC2). In critical applications, a high grade op amp might be substituted in place of the 741.

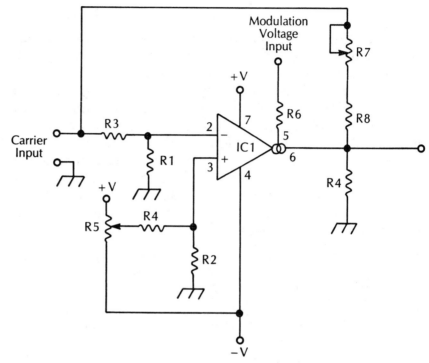

Fig. 4-18 Four-quadrant Multiplier

The complete parts list for this project is given in Table 4-13. In this circuit, the two inputs are labeled A and B. The A input corresponds to the carrier input of an amplitude modulator, and the B input is essentially the program input.

Table 4-13 Parts List for the Four-quadrant Multiplier project of Fig. 4-18.

Part	Description
IC1	CA3080 OTA IC
IC2	op amp (741, or similar)
R1	100K trimpot
R2	1 Megohm 5 percent 1/4-watt resistor
R3, R6	220 ohm 5 percent 1/4-watt resistor
R4	100K 5 percent 1/4-watt resistor
R5	39K 5 percent 1/4-watt resistor
R7, R10	22K 5 percent 1/4-watt resistor
R8	10K trimpot
R9	10K 5 percent 1/4-watt resistor

Notice that the A input signal is fed into the inverting input, pin 2 of the OTA, through resistor R4, while the B input voltage is converted into a current by resistor R7 and potentiometer R8, and then fed into the OTA's I-bias input (pin 5).

Either input voltage A or B may be positive or negative. The signals are effectively multiplied by one another in this circuit. The output polarity is correctly calculated, but inverted. If both inputs are of the same polarity, either both positive, or both negative, the output will be negative; while if the input signals are of opposite polarities, one positive and one negative, the output signal will be positive. Of course, this is just the opposite of true mathematical multiplication. The six opposite input combinations are shown in Table 4-14.

Table 4-14 Six Possible Input Combinations for the Four-quadrant Multiplier.

Input Polarity		Mathematical Result	Circuit Output Polarity
A	B		
+	+	+	−
+	−	−	+
−	+	−	+
−	−	+	−
0	?	0	0
?	0	0	0

Input conditions labeled "?" mean "don't care." The results will be the same, regardless of the polarity or value of the "?" input.

If the output polarity inversion should be a problem in your particular application, it would be easy enough to pass the output signal through an extra inverting stage (a unity gain inverting amplifier) so the output polarity matches the true mathematical results of multiplication.

The inputs to this circuit can safely handle signals up to 10 volts peak-to-peak (±5 volts). A simple resistive voltage divider network made up of resistors R4 and R6 drop the maximum input voltage down to the 10-millivolt maximum input level acceptable by the OTA (IC1).

Two calibration trimpots (R1 and R8) are included in this circuit. The calibration procedure is not terribly complicated. First, feed any suitable signal into the A input and ground the B input (input voltage of zero). Trimpot R1 is now adjusted for an output as close to zero as possible. Remember, the output of this circuit represents the results of multiplying the *A* signal by the *B* signal. Multiplying anything by zero, the *B* input, should give a result of zero.

Once trimpot R1 has been adjusted, reverse the input connections. That is, the external input signal should now be feeding the *B* input, and the *A* input should be grounded. Now, adjust trimpot R8. Again, the goal is to get an output as close to zero as possible.

Once these calibration procedures have been completed, controls R1 and R8 should be left alone. They should not be front panel control manual potentiometers.

The input sources to this circuit are assumed to have relatively low output impedances. If your application demands the use of high impedance signal sources, external buffering of the input signals may be required. As with the preceeding project, this circuit will work best if it is operated off of a full ±15-volt power supply.

A four-quadrant multiplier circuit like the one shown in Fig. 4-18 is useful in electronic music synthesis and analog computer applications.

PHASE-AMPLITUDE MODULATOR

Another amplitude modulation circuit is illustrated in Fig. 4-19. Besides modulating the signal amplitude, or level, this circuit also modulates the signal's *phase*. Very simply put, phase is when each cycle begins and ends. A suitable parts list for this project is given in Table 4-15.

Built around a CA3080 OTA IC, this circuit calls for the carrier input signal to be modulated to be applied to the inverting input, pin 2, while the modulating program signal is fed through resistor R6 to the I-bias input (pin 5). The output signal is tapped from the circuit across load resistor R9.

This circuit is basically quite similar to the straight amplitude modulation circuit shown in Fig. 4-15. The phase modulation effect is obtained by the addition of a feedback path through resistors R7 and R8 from the circuit's output back to the carrier input.

To calibrate the circuit, ground the program input (I-bias = 0). Adjust trimpot R7 so that the inverted signal at the OTA's

88 Switching and Modulation Projects

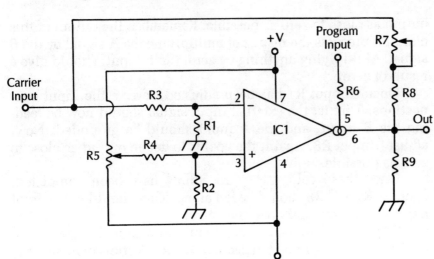

Fig. 4-19 Phase/Amplitude Modulator

Table 4-15 Parts List for the Phase/Amplitude Modulator project of Fig. 4-19.

Part	Description
IC1	CA3080 OTA
R1, R2	100 ohm 1/4-watt 5 percent resistor
R3, R8	56K 1/4-watt 5 percent resistor
R4	47K 1/4-watt 5 percent resistor
R5	100K trimpot
R6	33K 1/4-watt 5 percent resistor
R7	50K trimpot
R9	100K 1/4-watt 5 percent resistor

output is exactly balanced by the original uninverted carrier input signal. In other words, adjust potentiometer R7 for true zero output under these conditions.

Now, the phase of the output signal will be determined by the program (I-bias) input. If the instantaneous program input signal is positive, the OTA's output will be stronger than the original carrier input signal. If the program input signal is at its maximum positive value, $V+$, the output signal will be inverted.

Likewise, if the program input is at its maximum negative value, $V-$, the original carrier input signal through resistors R7 and R8 will overpower the OTA's output, and the output signal will be noninverted—in phase with the original carrier input signal.

To change the maximum gain of the circuit when the program input is at $V+$ or $V-$, change the values of resistors R3 and R8. The circuit gain is doubled if these resistor values are both halved. Increasing these resistances will decrease the gain of the OTA.

This circuit features a high output impedance. To drive a low impedance load, a buffer stage of some sort will be required. Whether or not an output buffer stage is necessary will depend on the application intended.

❖ 5
Signal-Generating Projects

GENERALLY SPEAKING, AN AMPLIFIER CIRCUIT ACCEPTS AN EXTERNAL input signal, which is then modified by the time it reaches the output. Under certain circumstances, however, an amplifier circuit can generate a signal of its own with no external input. This is due to positive feedback.

Almost everyone has experienced the unpleasant effects of acoustic feedback with a public address system. Some of the amplified sound from the loudspeakers is picked up by the microphone and repeatedly reamplified. The result is a loud, piercing howl or squeal.

Positive feedback in an electronic circuit is quite simple. Some of the output signal is returned to the input for reamplification. The output signal and the original input signal are in phase with one another, so they are effectively added together.

Unintentional and uncontrolled positive feedback is obviously undesirable. When an amplifier starts to oscillate, that is pretty much the ultimate in distortion. But controlled feedback and the resulting oscillation effect can often be useful.

The projects in this chapter demonstrate how transconductance operational amplifiers, such as the CA3080 and the LM13600 can be used in oscillator and other signal generation circuits.

SQUARE-WAVE GENERATOR

Figure 5-1 shows a practical square-wave generator circuit built around a CA3080 OTA IC. Another name for this type of circuit is an *astable multivibrator*.

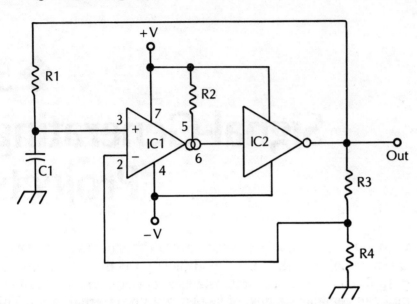

Fig. 5-1 Square-Wave Generator

A square wave has just two output conditions: HIGH and LOW. Theroretically, the signal level switches back and forth between these two states instantaneously, with no transition time, as illustrated in Fig. 5-2. In a practical circuit there is always a finite amount of transition time between the two extreme states, but this transition time, or slew, can be considered negligible for most uses.

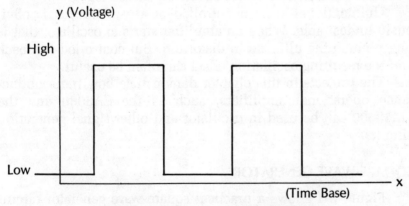

Fig. 5-2 A square wave switches back and forth between two discrete levels.

Square-Wave Generator   93

A suggested parts list for this project appears as Table 5-1. It is strongly suggested that you try breadboarding this circuit, and experiment with alternative component values, especially capacitor C1 and resistors R1, R3, and R4. These component values determine the frequency of the output signal generated by the circuit. In this circuit, the supply voltages should not exceed ±9 volts.

Table 5-1 Parts List for the Square-Wave Generator project of Fig. 5-1.

Part	Description
IC1	CA3080 OTA IC
IC2	CMOS inverter (CD4049, or similar)
C1	1000 pF capacitor
R1	47K 5 percent 1/4-watt resistor
R2	1 Megohm 5 percent 1/4-watt resistor
R3, R4	100K 5 percent 1/4-watt resistor

The output signal switches between a level just below the positive supply voltage (HIGH) and a level just above the negative supply voltage (LOW).

When the output goes HIGH, capacitor C1 starts to charge through resistor R1. At some point, the voltage stored in the capacitor will exceed the circuit's positive reference voltage.

This reference voltage is defined by the voltage divider network made up of resistors R3 and R4. The exact resistances are not important here, only the ratio between them. If these two resistors are given equal values as suggested in the parts list, the reference voltage will be equal to one-half the present output voltage. When the output is HIGH, the reference voltage is 4.5 volts.

Once capacitor C1 has charged to the positive reference voltage, the circuit is forced to reverse states and the output goes LOW. Now a negative reference voltage is tapped off between resistor R3 and resistor R4. If these resistors are equal, the reference voltage will be one-half the negative supply voltage, or −4.5 volts.

Capacitor C1 starts to discharge through resistor R1 until it reaches the negative reference voltage. This forces the circuit to reverse states again, and the output goes HIGH once more. This cyclic pattern continues indefinitely, as long as power is applied to the circuit.

Notice that the I-bias current, and thus the OTA's transconductance, is set by resistor R2, which has a fairly large value. The parts list recommends 1 megohm. This results in a rather low I-bias current. Because of the feedback in this circuit, high gain is not required. Remember that the CA3080's overall current gain is twice the I-bias value, so this circuit features a very low power-consumption rating. Depending on the CMOS inverter (IC2) used, the circuit may consume as little as a hundred microamps or less.

VARIABLE DUTY-CYCLE RECTANGULAR-WAVE GENERATOR

A square wave has a *duty cycle* of 1:2. The duty cycle is a measurement of how much of each complete cycle is in the HIGH state. A square wave is a special case of a rectangular wave. It is a symmetrical rectangular wave. That is, the LOW time exactly equals the HIGH time.

Rectangular waves may have other duty cycles, such as 1:3 and 1:4, as illustrated in Fig. 5-3. The duty cycle figure also defines the harmonic content of the waveform. A square wave,

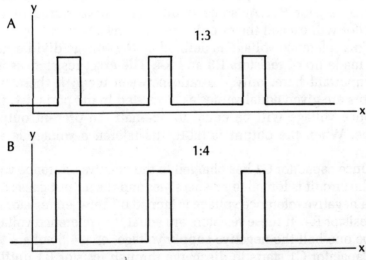

Fig. 5-3 Different rectangular waves may have different duty cycles:
A 1:3
B 1:4

with a duty of 1:2, has all harmonics except those which are evenly divisible by two. A square wave is comprised of the following frequency components:

- Fundamental
- Third harmonic
- Fifth harmonic
- Seventh harmonic
- Ninth harmonic,

and so forth.

A rectangular wave with a duty cycle of 1:3 contains all of the harmonics, except for those which are exact multiples of three:

- Fundamental
- Second harmonic
- Fourth harmonic
- Fifth harmonic
- Seventh harmonic
- Eighth harmonic
- Tenth harmonic,

and so forth.

If a rectangular wave has a duty cycle of 1:4, every fourth harmonic will be missing from the output waveform:

- Fundamental
- Second harmonic
- Third harmonic
- Fifth harmonic
- Sixth harmonic
- Seventh harmonic
- Ninth harmonic
- Tenth harmonic,

and so forth. Different applications may well call for rectangular waves of varying duty cycles.

The circuit illustrated in Fig. 5-4 permits manual control over the duty cycle of the output waveform. A suitable parts list for this project is given in Table 5-2.

96 Signal-Generating Projects

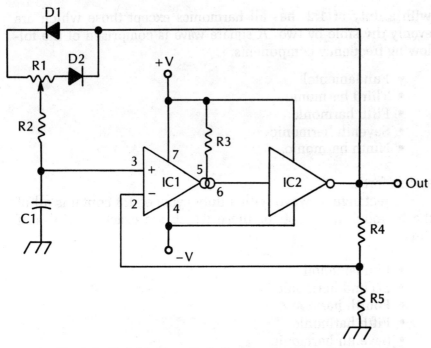

Fig. 5-4 Variable Duty-Cycle Rectangular-Wave Generator

**Table 5-2 Parts List for the
Variable Duty-Cycle Rectangular-Wave
Generator project of Fig. 5-4.**

Part	Description
IC1	CA3080 OTA IC
IC2	CMOS inverter (CD4049, or similar)
D1, D2	diode (1N914, or similar)
C1	1000 pF capacitor
R1	100K potentiometer
R2	10K 5 percent 1/4-watt resistor
R3	1 Megohm 5 percent 1/4-watt resistor
R4, R5	100K 5 percent 1/4-watt resistor

Notice how similar this circuit is to the one of Fig. 5-1. The main differences are the addition of the two diodes, D1 and D2, and the duty-control potentiometer, R1.

Capacitor C1 is charged and discharged through resistor R2 and half of potentiometer R1, depending on the position of the slider. During each half of the cycle, one of the diodes is conducting and the other half blocks current flow.

If potentiometer R1 is set to the exact middle of its range, the same resistance will be seen by the circuit during both halves of the cycle, so the result will be a symmetrical square wave. For any other setting of potentiometer R1, different resistances will be seen during the LOW and HIGH portions of the cycle. The LOW time will be either longer or shorter than the HIGH time, resulting in a nonsymmetrical rectangular waveform at the circuit's output.

As with the previous square-wave generator circuit, the supply voltages for this circuit should not exceed ±9 volts. The output signals will essentially switch between the V+ (HIGH) and V− (LOW) supply voltages.

Again, notice that the I-bias current, and thus the OTA's transconductance, is set by resistor R3, which has a fairly large value. The parts list recommends 1 megohm, which results in a rather low I-bias current. Because of the feedback in this circuit, high gain is not required. Remember that the CA3080's overall current gain is just twice the I-bias value, so this circuit features a very low power consumption rating. Depending on the CMOS inverter (IC2) used, the circuit may consume as little as a hundred microamps or less.

Depending on the specific application intended, potentiometer R1 may be either a front panel manual control, or a calibration type screwdriver adjust trimpot.

You should be aware of one important limitation involved with this project. Changing the duty cycle in this circuit will alter the frequency of the output waveform.

VCO

Our next project is a *voltage-controlled oscillator* (VCO). This is a signal generating circuit in which the output frequency is determined by a controlling voltage input signal. VCO applications include electronic music, automatic tuning, and other automation systems.

A practical VCO circuit uses transconductance operational amplifiers as illustrated in Fig. 5-5. A suitable parts list for this project is given in Table 5-3.

98 Signal-Generating Projects

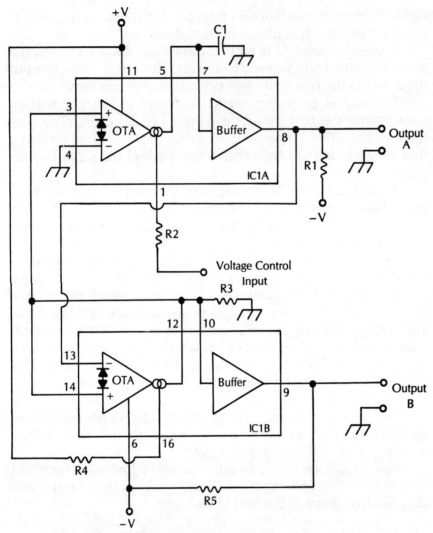

Fig. 5-5 Voltage-controlled Oscillator

Notice that this project uses both halves of an LM13600 dual OTA IC. The two sections are shown separately in the schematic diagram for a more convenient component layout and greater clarity.

This VCO circuit has two separate outputs. One of these outputs is a square wave, while the other output generates a triangular wave. Both output waveforms are simultaneously available. Both output signals will always be at the same frequency. A square wave and a triangular wave have the same basic harmonic

Table 5-3 Parts List for the
VCO project of Fig. 5-5.

Part	Description
IC1	LM13600 dual OTA
C1	400 pF capacitor
R1, R5	10K 1/4-watt 5 percent resistor
R2	22K 1/4-watt 5 percent resistor
R3	4.7K 1/4-watt 5 percent resistor
R4	47K 1/4-watt 5 percent resistor

content. The difference between these two waveforms lies in the relative amplitude of the harmonics. The harmonics in a triangular wave are much weaker with respect to the fundamental frequency than in a square wave.

In the operation of this circuit, capacitor C1 is first charged positively, then discharged and charged negatively, then discharged again, and the cycle is repeated. The fundamentals of this portion of the circuit are not entirely dissimilar to those of the square-wave generator project described earlier in this chapter.

When capacitor C1 reaches its full negative charge, the square-wave output goes HIGH. A positive reference voltage is then developed across resistor R3. This reference is fed back into the noninverting inputs of the two amplifiers, IC1A and IC1B.

IC1A starts to generate a positive output current. This positive output current from IC1A will be equal to the I-bias current at pin 1. The output current from IC1A starts to discharge, then charge capacitor C1 in the positive direction, generating a fairly linear ramp voltage, which is buffered and then fed into the inverting input of IC1B. At some point, this steadily increasing voltage at the inverting input of IC1B will exceed the positive reference voltage at the noninverting input. This forces the output of this OTA to switch states and go LOW.

Now a negative reference voltage is placed across resistor R3. This reference is fed back into the noninverting inputs of both OTAs, IC1A and IC1B. IC1A starts to generate a negative output current. This negative output current from IC1A starts to discharge, then charge capacitor C1 in the negative direction, generating a fairly linear ramp voltage in the opposite direction, as before.

This descending ramp voltage is buffered, and then fed into the inverting input of IC1B. At some point, this steadily decreasing voltage at the inverting input of IC1B will drop below the negative reference voltage at the noninverting input. This forces the output of this OTA to switch states once more and go HIGH.

The entire two-part cycle described above is continuously repeated as long as power is applied to the circuit. The frequency of the output signal is a result of how fast capacitor C1 can be charged and discharged. This parameter is controlled by the values of capacitor C1, resistor R3, and the I-bias current fed to IC1A (pin 1).

Increasing the I-bias current increases the output frequency, and vice versa. The I-bias current for IC1A, in turn, is determined by the value of resistor R2, and the applied control voltage. Once the circuit has been constructed, the component values of R2, R3, and C1 will be constant, and the output frequency will be controlled directly by the signal applied to the control voltage input.

The larger the control voltage fed into the circuit, the higher the frequency of the output signal. Using the component values suggested in the parts list (Table 5-3), the minimum output frequency for this circuit is about 200 Hz, and it can go up to nearly 200 kHz.

RANDOM MUSIC MAKER

I've been an electronic music enthusiast for a long time. I particularly enjoy automated music projects—that is, instruments which "play" themselves. A very interesting circuit of this type is shown in Fig. 5-6. A suitable parts list for this project appears as Table 5-4. Experimentation with alternative component values in this circuit is encouraged.

This is the most complex circuit in this book, using four ICs. IC1 is a 567 phase-locked loop (PLL) IC, which is being used here as a VCO.

IC3 is a CA3080 OTA IC, and IC4 is a standard op amp IC. A high-grade, low-noise unit is recommended for this application. A 741 will work, but the CA3140 called for in the parts list will offer better performance.

IC4 is a 555 timer, which is wired as a simple astable multivibrator, or rectangular wave generator.

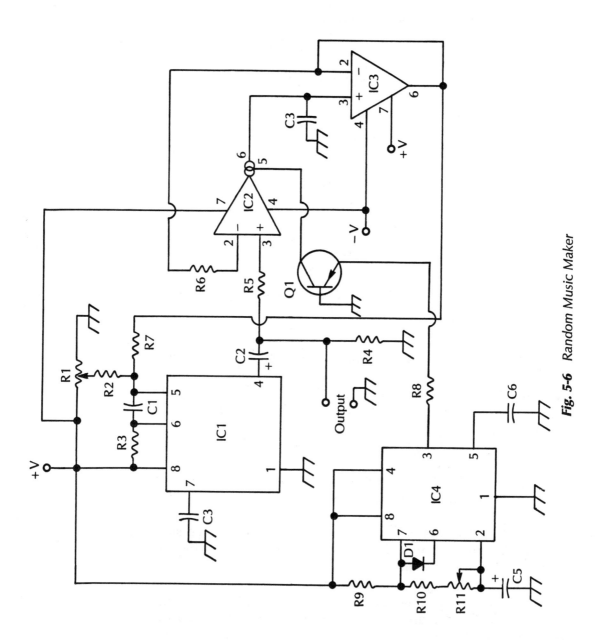

Fig. 5-6 Random Music Maker

Table 5-4 Parts List for the Random Music Maker project of Fig. 5-6.

Part	Description
D1	1N4148
Q1	PNP 2N2907 (or similar)
IC1	567 PLL
IC2	CA3080 OTA
IC3	op amp (CA3140, or similar)
IC4	555 timer
C1, C6	0.001 µF capacitor
C2	1 µF 25 volt electrolytic capacitor
C3	0.05 µF capacitor
C4	0.1 µF capacitor
C5	4.7 µF 25 volt electrolytic capacitor
R1	10K potentiometer
R2, R10	4.7K 1/4-watt 5 percent resistor
R3, R5, R6, R9	10K 1/4-watt 5 percent resistor
R4	100K 1/4-watt 5 percent resistor
R7	2.2K 1/4-watt 5 percent resistor
R8	22K 1/4-watt 5 percent resistor
R11	500K potentiometer

IC2 and IC3 and their associated components form a Sample-and-Hold circuit, as described in chapter 4. In fact, this is the same circuit we worked with in that chapter. In this project, we simply have a built-in input signal source (IC1) and trigger pulse source (IC4).

The signal generator built around IC4 is set for a relatively low, subaudible frequency. Each time the output of this oscillator switches from LOW to HIGH, it triggers the Sample-and-Hold circuitry to sample the instantaneous signal voltage from the VCO (IC1).

The output of IC1 is a triangular wave, so the voltage sampled by IC3 and IC4 will depend on exactly when during the cycle the circuit is triggered.

To make things even more interesting, feedback is used from the output of the Sample-and-Hold to the control input of the VCO. The VCO's present signal frequency is determined by the last sampled value. The results of this little trick make the output signal very unpredictable. Adjusting potentiometer R1 determines the overall frequency range of the VCO.

The only other manual control in this circuit is potentiometer R11. The resistance of R11 sets the sampling rate, or the frequency of the trigger pulses. If a very low frequency is used for the trigger pulses, the output will be a pseudorandom series of discrete voltages, which can be used to drive an external VCO or other voltage-controlled circuit.

Increasing the trigger frequency into the audible range allows this circuit to be used by itself as a complex tone generator. At high trigger frequencies, the result will be a harsh tone with a continuously varying timbre.

At moderate trigger frequencies, a pseudorandom series of discrete tones, or "notes" will be heard. The circuit will play a little "tune" which it "makes up" as it goes along. Of course, the results probably won't sound much like real music. They will sound more like a robot or computer out of a science fiction film.

Even though it's really not all that musical, this random music maker circuit is fascinating to experiment with. It's great for use in customized children's toys too.

By all means, experiment with alternative component values in this project. I particularly recommend trying different values for capacitors C1, C2, and C5, and resistors R3, R4, R7, and R9.

❖ 6
Miscellaneous OTA Projects

THIS CHAPTER FEATURES A POTPOURRI OF CIRCUITS USING transconductance operational amplifiers. Many of these projects don't have much to do with one another, but they are of interest for various reasons. Rather than break up the book into a number of short chapters of one or two projects each, I have collected these circuits into this "Miscellaneous Chapter."

HARMONICS REMOVER

Most ac waveforms are made up of numerous frequency components, including the fundamental, or nominal frequency, and some combination of harmonics—whole number multiples of the fundamental frequency. The one exception is the sine wave, which consists of the fundamental, with no harmonic content at all.

The circuit shown in Fig. 6-1 deletes the harmonics from an input signal, leaving theoretically a sine wave at the output. A suitable parts list for this project is given in Table 6-1.

If the input signal is at a known, fixed frequency, a low-pass filter could be used to attenuate the harmonics, since the harmonics are, by definition, all at higher frequencies than the fundamental. However, it can be difficult to design a filter circuit with a steep enough cutoff slope. Also, if the signal frequency changes, so will the harmonic content of the output signal, because the filter's cutoff frequency is fixed. An exception is a *voltage-controlled filter (VCF)*. Later in this chapter we will work with some VCF circuits.

106 Miscellaneous OTA Projects

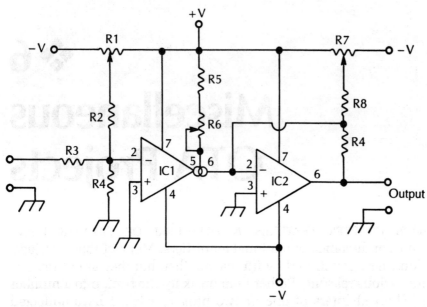

Fig. 6-1 Harmonics Remover

**Table 6-1 Parts List for
the Harmonics Remover project of Fig. 6-1.**

Part	Description
IC1	CA3080 OTA IC
IC2	op amp (741, or similar)
R1, R7	100K trimpot
R2, R8	1 Megohm 5 percent ¼-watt resistor
R3	68K 5 percent ¼-watt resistor
R4	1K 5 percent ¼-watt resistor
R5	220K 5 percent ¼-watt resistor
R6	50K trimpot
R9	39K 5 percent ¼-watt resistor

The Harmonic Remover circuit of Fig. 6-1 is not frequency dependent. This is because there is virtually no reactance (ac resistance) in this circuit. The circuit does not include any reactive (frequency-sensitive) components—capacitors or inductors.

This circuit is a case of deliberately doing something "wrong" to obtain a desired result. As you should recall from the earlier chapters in this book, the inverting and noninverting inputs of the CA3080 OTA IC should not be fed signals higher than 10 mV, or distortion will result. In this circuit, we are deliberately overdriving the OTA's inputs to generate controlled distortion.

The input signal is fed into the OTA's inverting input, pin 2, through resistor R3. Assuming the original input signal has an amplitude of 10 volts peak-to-peak, ±5 volts, the input resistor will drop the signal level down to about 150 mV peak-to-peak. This is well above the CA3080's 10 mV limit.

The OTA is overdriven by this excessive (but not dangerously so) input signal. The gain, or transconductance of the amplifier is determined by the current flowing through resistor R5 and potentiometer R6. This potentiometer should be a screwdriver adjust trimpot, not a manual front panel control. We'll get to the use of this potentiometer in a moment.

The two other potentiometers in the circuit, R1 and R7, should also be screwdriver adjust trimpots. These three controls interact somewhat, and must be carefully adjusted during the calibration procedure.

During calibration, the circuit's output signal is monitored, ideally with an oscilloscope. First, the input is grounded for a 0-volt input. Potentiometer R7 is adjusted to trim the output offset. Adjust this control for a true 0 output, or as close as possible, with the circuit's input grounded.

Next, remove the short between the circuit input and ground. Apply a suitable 10-volt peak-to-peak input signal to the input of the circuit. Watch the waveform on the oscilloscope very carefully during the following adjustments. First, adjust potentiometer R1 to make the output waveform as symmetrical as possible. If the waveform is perfectly symmetrical, it will contain no even harmonics.

Now, it's just a matter of getting rid of the odd harmonics, which is the function of R6. This potentiometer is carefully adjusted for the minimum amplitude of the odd harmonics.

You may have to alternate between R1 (even harmonic adjust) and R6 (odd harmonic adjust) several times to get the absolute minimum of harmonic content in the output signal.

This circuit works best if the original input signal's harmonic content is not terribly strong. A triangular wave would be an excellent choice.

108 Miscellaneous OTA Projects

With careful calibration, a very good approximation of a pure sine wave can be achieved at the output of the circuit. Typically, the distortion, or nonfundamental content in the output signal can be adjusted as low as 2 percent to 5 percent. This is better than many sine wave oscillator circuits.

VOLTAGE-CONTROLLED RESISTANCE

In many applications, especially those involving remote control, or automation, it is desirable to have some circuit parameter that can be adjusted by purely electrical, rather than manual means.

The circuit shown in Fig. 6-2 converts an input voltage into a proportional output resistance. Of course, due to Ohm's Law,

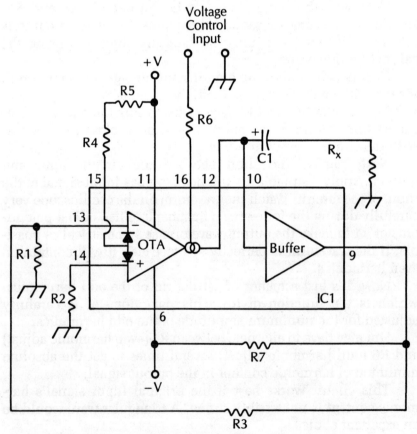

Fig. 6-2 Voltage-controlled Resistance Circuit

this circuit could also function as a voltage controlled current, or even, odd as it sounds, a voltage controlled voltage. A suitable parts list for this project appears in Table 6-2.

Table 6-2 Parts List for the Voltage-controlled Resistance project of Fig. 6-2.

Part	Description
IC1	LM13600 dual OTA IC
C1	2 µF 35 Volt electrolytic capacitor
R1, R2	470 ohm 5 percent 1/4-watt resistor
R3	100K 5 percent 1/4-watt resistor
R4	12K 5 percent 1/4-watt resistor
R5	3.3K 5 percent 1/4-watt resistor
R6	27K 5 percent 1/4-watt resistor
R7	10K 5 percent 1/4-watt resistor

The output resistance is measured across the two points labeled R_x in the diagram. These two points can be wired into almost any ac circuit, just as if a standard fixed resistor were being used. The only real restriction is that one end of the voltage controlled resistor must be at ground potential. Also, the voltage controlled resistor circuit will only function properly if the signal across the R_x terminals is an ac waveform.

The effective ac resistance across the R_x terminals is determined by the values of resistors R1 and R3, along with the gain, or transconductance of the OTA. The transconductance is determined by the I-bias current developed across resistor R6, which is proportional to the level of the voltage control input. The effective R_x resistance can be found with this formula:

$$R_x = R_5 / (20 \times R_1)$$

Using the component values suggested in the **parts list** (R_1 = 470 ohms, R_3 = 100K, and R_6 = 27K), the effective R_x resistance can be varied over a 10K to 10 megohm range. The larger the control voltage, the larger the I-bias current will be, due to Ohm's Law:

$$I = E/R$$

The resistance in this case, R_6, is assumed to be a constant value. Increasing the I-bias current in this circuit, lowers the effective ac resistance across the R_x terminals.

In principle, the OTA in this circuit is generating an output current which is proportional to the I-bias current and the ac voltage across the R_x terminals. Ohm's Law comes into play once again, allowing the OTA's output current to simulate a resistance:

$$R = E/I$$

E, in this case, is the ac voltage applied across the R_x terminals, and I is the OTA's output current. Assuming the voltage is held constant, the current, I, will vary in proportion to changes in the I-bias current. The effective resistance will then be varied in step with the output current.

This tricky little circuit is quite unusual and specialized, but it can come in very handy.

PRECISION CURRENT SOURCE

Generally speaking, current flow in a circuit is controlled by Ohm's Law:

$$I = E/R$$

When one circuit drives another, the load circuit essentially acts as a resistance, typically designated "R1." In operation, many factors can cause the input impedance of the load to change—either increasing or decreasing. This will normally cause a change in the current drawn from the source circuit. If the load resistance increases, the current drawn decreases and vice versa.

For some uses, we will need a precise current that remains constant, regardless of fluctuations in the load resistance or impedance.

The circuit shown in Fig. 6-3 is a handy precision current source. A suitable parts list for this project is given in Table 6-3. The boxed resistor marked R1 in the schematic diagram represents the load impedance.

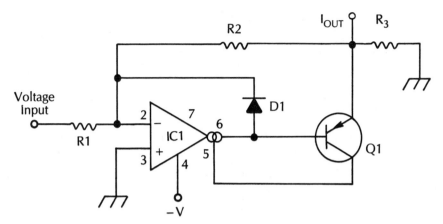

Fig. 6-3 Precision Current Source

**Table 6-3 Parts List for the
Precision Current Source project of Fig. 6-3.**

Part	Description
IC1	CA3080 OTA
D1	diode (1N914, 1N4148, or similar)
R1, R2	100K 1/4-watt 5 percent resistor
R3	39K 1/4-watt 5 percent resistor

This circuit is not particularly critical in its component requirements. Transistor Q1 can be almost any low-power transistor of the PNP type. Diode D1 is a simple signal switching diode, such as the 1N914, or the 1N4148. The voltage input to this circuit is converted into a precise and steady output current.

LOW-PASS VCF

Filters are used in a great many different applications. A filter is a frequency sensitive circuit. It is essentially a modified amplifier with a deliberately nonlinear frequency response.

A filter passes some frequency components in the input signal through to its output, while other frequency components are blocked, or significantly attenuated.

112 Miscellaneous OTA Projects

There are four basic filter types:

- Low-pass
- High-pass
- Band-pass
- Band-reject

A *low-pass filter* circuit has a specific cutoff frequency. Everything below this cutoff frequency—low frequency components—is passed, while higher frequencies above the cutoff frequency are blocked. A frequency response graph for an ideal low-pass filter is illustrated in Fig. 6-4.

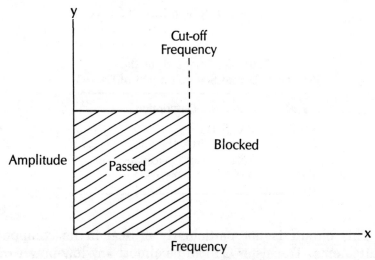

Fig. 6-4 *A low-pass filter blocks high frequencies, while passing low frequencies.*

In a practical filter circuit, there is an intermediate region between fully passed and fully blocked frequency components. Any frequency components close to the cutoff frequency are partially attenuated, as illustrated in Fig. 6-5. For most purposes, the steeper the cutoff slope, the better the filter.

A *high-pass filter* is just the opposite of a low-pass filter. As shown in the frequency response graph of Fig. 6-6, all frequency components above the cutoff frequency are passed, while lower frequency components are blocked by the circuit.

A *band-pass filter* passes a specific band of frequencies, while rejecting any frequency components that fall outside,

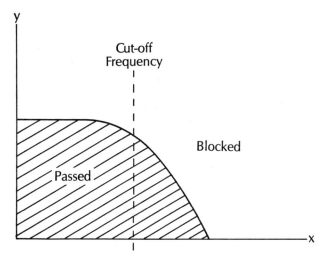

Fig. 6-5 *A practical filter has a noninfinite cutoff slope.*

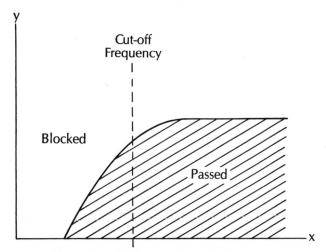

Fig. 6-6 *A high-pass filter blocks high frequencies, while passing low frequencies.*

either above or below the specified band. The frequency response of a typical band-pass filter is shown in Fig. 6-7.

A *band-reject filter* is the opposite of a band-pass filter. A band-reject filter passes all frequency components except those within the specified band. Because of the appearance of the frequency response graph, as illustrated in Fig. 6-8, this type of circuit is often known as a *notch filter*.

114 Miscellaneous OTA Projects

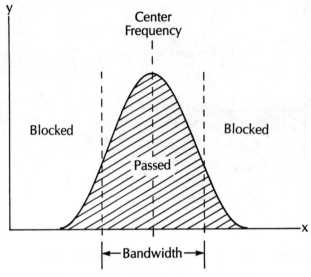

Fig. 6-7 A band-pass filter passes only those frequencies which fall in a specific band.

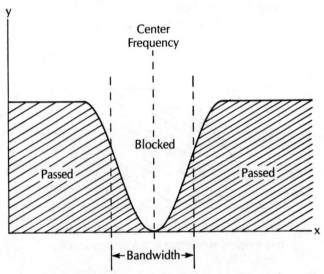

Fig. 6-8 A band-reject filter passes all frequencies except for those which fall within a specified band.

In this project we will be working with a low-pass filter. A high-pass filter circuit will be presented in the following project. We will not be dealing with band-pass filter or band-reject filters in this section.

Simple filter circuits have a fixed cutoff frequency, determined by specific component values within the circuit. This is fine for some applications, but in other uses, where the input signal frequency might change with time, a fixed filter could cause operational problems. The harmonic content of the output signal will vary with the signal frequency.

This is best illustrated with a specific example. In the following discussion, we will assume that the input signal is always a square wave. A square wave is comprised of the fundamental frequency and all of the odd harmonics:

- Fundamental
- Third harmonic
- Fifth harmonic
- Seventh harmonic
- Ninth harmonic
- Eleventh harmonic
- Thirteenth harmonic,

and so on.

For convenience, we will ignore all frequency components above the thirteenth harmonic in the following discussion. The upper harmonics are usually very weak, so they can reasonably be ignored.

Let's say we are feeding this square-wave signal through a fixed low-pass filter circuit with a cutoff frequency of 560 Hz. For simplicity, we will assume that this is an impossibly ideal filter with an infinite cutoff slope. A frequency component at 559 Hz will be fully passed, and a frequency component at 561 Hz will be completely blocked.

If the input signal has a frequency of 100 Hz, it will consist of the following frequency components:

$$100 \text{ Hz}$$
$$300 \text{ Hz}$$
$$500 \text{ Hz}$$
$$700 \text{ Hz}$$
$$900 \text{ Hz}$$
$$1100 \text{ Hz}$$
$$1300 \text{ Hz}$$

At the output of the filter, the signal will be stripped to three frequency components:

        100 Hz
        300 Hz
        500 Hz

Now, let's see what happens when we raise the input signal frequency to 200 Hz. The harmonic makeup of the original input signal will be as follows:

        200 Hz
        600 Hz
        1000 Hz
        1400 Hz
        1800 Hz
        2200 Hz
        2600 Hz

Passing this signal through our theoretical 560 Hz filter will result in all of the harmonics being blocked. The output signal will be nothing more than the fundamental—200 Hz.

If we raise the input signal frequency to 600 Hz, or anything above the filter's cutoff frequency, the entire signal will be blocked, and the output will consist of nothing at all.

Next, let's see what happens when we reduce the frequency of the input signal. If the input signal has a frequency of 50 Hz, its harmonic makeup can be broken down as follows:

        50 Hz
        150 Hz
        250 Hz
        350 Hz
        450 Hz
        550 Hz
        650 Hz

Almost the entire signal will get through the filter in our example. The output signal will consist of these frequency components:

> 50 Hz
> 150 Hz
> 250 Hz
> 350 Hz
> 450 Hz
> 550 Hz

The harmonic content of the output signal will vary with changes in the frequency of the input signal.

In some applications, a suitable solution to this problem is to use a *voltage-controlled filter (VCF)*. In this type of circuit, the cutoff frequency varies in proportion to a control voltage input. This approach is often used in electronic music systems, where the input signal is generated by a VCO (see chapter 5). By using the same control voltage for both the VCO and the VCF, the harmonic content will remain stable, regardless of the signal frequency.

A practical low-pass VCF circuit built around one-half of an LM13600 dual OTA IC is shown in Fig. 6-9. A suitable parts list for this project is given as Table 6-4.

Using the component values suggested in the parts list, the cutoff frequency can cover a range extending from about 50 Hz to a little under 50 kHz. This wide range is more than adequate for most practical audio uses.

To change the available range of cutoff frequencies, experiment with other values for resistor R5 and capacitor C2. For any given set of component values, the exact cutoff frequency is determined by the present value of the I-bias current. The I-bias current is developed by feeding a control voltage through resistor R3.

Capacitor C1 blocks any dc component in the signal input. It is not practical to try to filter a dc signal. The input signal to this VCF circuit is fed into the OTA's (IC1's) noninverting input through capacitor C1 and a voltage divider network made up of

118 Miscellaneous OTA Projects

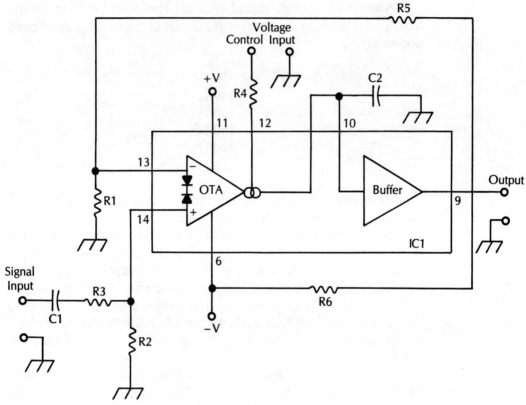

Fig. 6-9 Low-pass VCF

Table 6-4 Parts List for the Low-pass VCF project of Fig. 6-9.

Part	Description
IC1	LM13600 dual OTA
C1	0.5 uF capacitor
C2	180 pF capacitor
R1, R2	220 ohm 1/4-watt 5 percent resistor
R3	120K 1/4-watt 5 percent resistor
R4	22K 1/4-watt 5 percent resistor
R5	100K 1/4-watt 5 percent resistor
R6	10K 1/4-watt 5 percent resistor

resistors R2 and R3. These resistors reduce the input signal level to a level that can be handled by the OTA. Using the component values suggested in the parts list, the original input signal may vary over approximately a ±5-volt range.

Since the noninverting input is being used here, the OTA's output is in phase with the input. The OTA's output, pin 12, is coupled to the buffer input, pin 10, and a capacitor, which shunts high-frequency components to ground. Low-frequency components are blocked by the capacitor, and are forced to enter the output buffer.

Part of the output signal, at pin 9, is fed back into the OTA's inverting input through a voltage divider network made up of resistors R5 and R1. The feedback signal (inverting input) cancels part of the original input signal (noninverting input).

At low frequencies, capacitor C2 has a very high impedance, so the capacitor becomes fully charged by the OTA's current output. The result of this is that the circuit functions very much like a unity gain voltage follower, or like a simple buffer amplifier.

As the signal frequency is increased, however, the impedance of capacitor C2 decreases. The capacitor can no longer fully charge up from the OTA's current output, so the signal amplitude is attenuated (drops below unity gain). The amount of attenuation is directly proportional to how high the signal frequency is with respect to the filter's cutoff frequency as determined by the values of resistor R5, capacitor C2, and the I-bias current. The cutoff slope for this particular circuit is 6 dB per octave. That is, the signal level is attenuated 6 db for every doubling of the signal frequency.

For technical purposes, when discussing filters, the cutoff frequency is defined as the frequency point where the output amplitude has dropped 3 dB.

HIGH-PASS VCF

Figure 6-10 illustrates how half an LM13600 dual OTA IC can be used as a high-pass VCF. A suitable parts list for this project is given in Table 6-5. By all means, experiment with other component values.

Using the component values suggested in the parts list, the cutoff frequency can be adjusted by applying a suitable control voltage over a 6 Hz to 6 kHz range.

Remember, a high-pass filter passes all frequency components *above* the cutoff frequency on to the circuit output, while blocking the frequency components which are *lower* than the cutoff frequency.

120 Miscellaneous OTA Projects

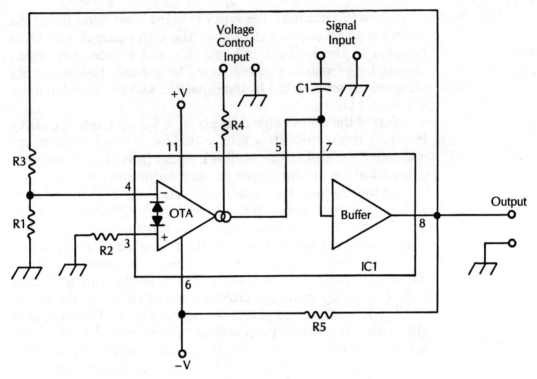

Fig. 6-10 High-pass VCF

Table 6-5 Parts List for the High-pass VCF project of Fig. 6-10.

Part	Description
IC1	LM13600 dual OTA
C1	5000 pF capacitor
R1, R2	1K ¼-watt 5 percent resistor
R3	100K ¼-watt 5 percent resistor
R4	22K ¼-watt 5 percent resistor
R5	10K ¼-watt 5 percent resistor

Notice that the input signal is fed directly into the buffer, pin 7—not to the inverting input (pin 4) or the noninverting input (pin 3) of the OTA. The linearizing diode input (pin 2) is not used in this circuit.

In the last few chapters we have looked at over two dozen practical circuits built around transconductance operational amplifiers. Beginning with the next chapter, we will start working with another specialized type of op amp: the *Norton amplifier*.

Part II
Norton Amplifiers

❖ 7
The Norton Operational Amplifier

ALL OPERATIONAL AMPLIFIERS, BY DEFINITION, ARE DIFFERENTIAL amplifiers. Any op amp has an inverting input and a noninverting input. The output is determined by the difference between these two input signals, multiplied by the amplifier's gain.

A conventional op amp, such as the 741, is a voltage differential amplifier. Both the input signals and the output signal are in the form of voltages.

A transconductance operational amplifier, or OTA, such as the CA3080, accepts input signals in the form of voltages, but its output is in the form of a current.

Another specialized type of operational amplifier is the *Norton amplifier*. In very simple terms, a Norton op amp is a current differential amplifier. The output is proportional to the difference between two input currents. The output from a Norton op amp is normally in the form of a voltage.

BASIC PRINCIPLES OF THE NORTON OP AMP

Although *Norton op amp* or *Norton amplifier* is the most commonly used terminology, this device is sometimes referred to as a *current-differencing amplifier (CDA)*. This alternate name is more descriptive.

The standard schematic symbol for a conventional op amp is shown again in Fig. 7-1. Compare this with the symbol for a Norton op amp, illustrated in Fig. 7-2. Notice the small arrow between the inverting input and the noninverting input. This arrow indicates that the device accepts input signals in the form of currents, rather than voltages.

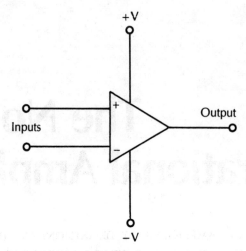

Fig. 7-1 *The standard schematic symbol for a conventional operational amplifier.*

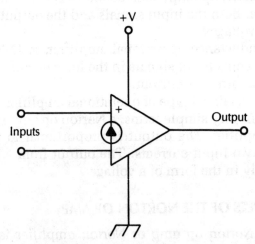

Fig. 7-2 *This schematic symbol is used to represent a Norton amplifier.*

Unlike most conventional op amp devices, a Norton amplifier does not require a dual-polarity power supply. A single positive supply voltage, referenced to circuit ground, can drive this device. This difference in power supply requirements can greatly reduce the cost and bulk of many circuits. It can be a particular advantage in portable applications.

Norton op amp circuits are otherwise very similar to conventional op amp circuits. Norton op amps often have many of the same uses as a standard op amp. In addition, the unique characteristics of the Norton amplifier make it suitable for some specialized applications of its own.

Norton op amp applications will be discussed further, later in this chapter, and a number of practical circuits using this device will be presented in chapters 9 through 11.

BIASING

To function correctly, a Norton op amp must be properly biased. In most circuits, biasing is accomplished very simply by applying a fixed voltage through a single resistor, as shown in Fig. 7-3. This is a simple inverting amplifier circuit. The resistor we are interested in at this time is the one labeled R_b. The applied voltage used for biasing the Norton op amp is usually just the circuit's supply voltage $(V+)$.

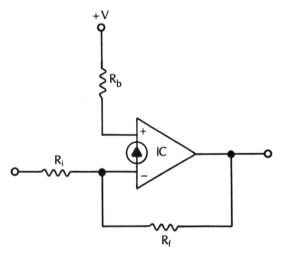

Fig. 7-3 A Norton amplifier must be properly biased.

Each of the input terminals in the Norton op amp is at a level equal to one diode drop above ground. Typically, this works out to approximately 0.6 volt. Because of this, bias resistor R_b should have a value double that of the circuit's feedback resistor (R_f). This biases the Norton op amp's steady-state (no input signal) output voltage to the midpoint of the circuit's supply voltage.

For example, if a Norton amplifier circuit is being run by a +9-volt power supply, the output voltage should be +4.5 volts when there is no input to the circuit.

If the bias resistor has a value of 100K, and the supply voltage is +9 volts, then Ohm's Law tells us that the bias current through the bias resistor is equal to:

$$\begin{aligned} I &= E/R \\ &= 9/100000 \\ &= 0.00009 \text{ ampere} \\ &= 0.09 \text{ mA} \\ &= 90 \text{ } \mu A \end{aligned}$$

INVERTING CIRCUITS

The internal circuitry of a Norton op amp includes negative feedback which forces both of the inputs, inverting and noninverting, to draw equal currents. This is very similar to the situation found in a conventional operational amplifier, except currents are substituted for voltages. The negative feedback in a standard op amp causes equal voltages to appear at the inverting and noninverting inputs.

In the Norton amplifier circuit of Fig. 7-3, the current drawn through the feedback resistor (R_f) must be equal to the current through the bias resistor (R_b) when the input is zero. Ohm's Law can again be used to select a suitable feedback resistor value so that the output voltage is at the midpoint of the supply voltage when the input is zero. In our example, half the supply voltage is 4.5 volts and the current flowing through the feedback resistor is 90 μA, so the desired resistance works out to a value of:

$$\begin{aligned} R &= E/I \\ &= 4.5/0.00009 \\ &= 50,000 \text{ ohms} \\ &= 50 \text{ K}\Omega \end{aligned}$$

In a practical circuit, a 47K resistor would probably be used in the feedback path, because it is the closest standard resistance value. This will cause some offset of the output voltage. In a precision application, a small trimpot would be added in the feedback path, as illustrated in Fig. 7-4. In our design example, a

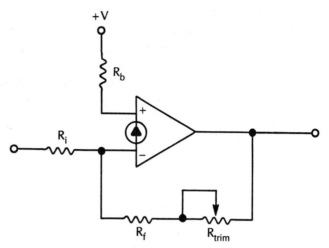

Fig. 7-4 *A small trimpot can be added in the feedback path to precisely bias a Norton amplifier.*

good value for the trimpot would be 10K, and the fixed feedback resistor, R_f, would have a value of 47K, or possibly 42K.

The output can now be precisely adjusted for exactly +4.5 volts when the input of the circuit is grounded (input = 0). The use of a calibration trimpot also compensates for any variations in the true, as opposed to nominal, values of any of the circuit components.

Now, if the input voltage applied to this circuit goes negative, the output voltage will drop below half the supply voltage. Likewise, a positive input voltage will cause the output voltage to be greater than one-half the supply voltage.

The input voltage is converted into a current by passing it through the input resistor, R_i. This gives us three currents of interest in the circuit:

I_b The bias current through resistor R_b
I_f The feedback current through resistor R_f
I_i The input current through resistor R_i

The internal design of the Norton op amp forces the current at the inverting input to be equal to the current at the noninverting input. In algebraic terms:

$$I_b = I_f + I_i$$

Since the input currents seen by the Norton amplifier must be equal, any current fluctuation through the input resistor, R_i, must be matched by an equal current change through the feedback resistor, R_f. The feedback current fluctuation will always have an opposite polarity from the input current fluctuation.

In other words, the current flowing through the feedback resistor, R_f, is always equal to, but polarity reversed from, the current flowing through the input resistor, R_i. That is:

$$I_i = I_f$$

Ohm's Law lets us rewrite this equation as follows:

$$V_i/R_i = -V_o/R_f$$

Notice that the negative sign indicates that the polarity is inverted at the output.

Amplifier gain is simply a comparison of the input and output voltages, so in this circuit the gain is equal to:

$$G = -V_o/V_i$$
$$= -R_f/R_i$$

Notice that this gain equation is exactly the same as the gain equation for an inverting amplifier circuit using a conventional op amp (see Fig. 7-5). Norton op amps have many of the same uses as conventional operational amplifiers.

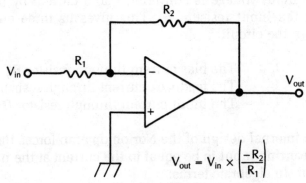

Fig. 7-5 An inverting Norton amplifier uses the same gain equation as a conventional inverting amplifier.

NONINVERTING CIRCUITS

Converting the Norton amplifier circuit of Fig. 7-3 from an inverting amplifier to a noninverting amplifier is quite simple. It's just a matter of moving the input resistor, R_i, from the inverting input to the noninverting input, as shown in Fig. 7-6.

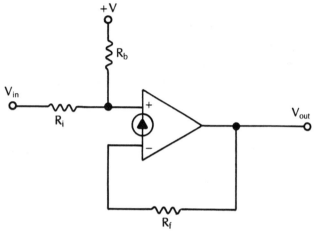

Fig. 7-6 *A noninverting Norton Amplifier*

For comparison purposes, a noninverting amplifier circuit built around a conventional op amp device is illustrated in Fig. 7-7.

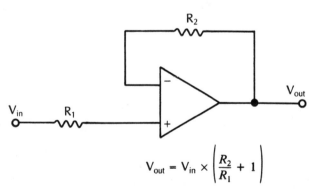

$$V_{out} = V_{in} \times \left(\frac{R_2}{R_1} + 1\right)$$

Fig. 7-7 *A conventional noninverting amplifier circuit*

The noninverting circuit of Fig. 7-6 functions very like the inverting circuit of Fig. 7-3. Once again, the current flowing through the feedback resistor, R_f, must always equal the current flowing through the input resistor, R_i. The big difference here is that the input current, I_i, has the same polarity as the feedback current, I_f. There is no inversion of the signal in this circuit.

The gain in this circuit can be found with this simple formula:

$$G = V_o/V_i$$

Notice that there is no negative sign in this equation, because the signal polarity is not inverted at the output.

Because of the circuit equalities described above, the gain equation can be easily rewritten in terms of resistance. That is:

$$G = R_f/R_i$$

The gain equation for a noninverting circuit using a conventional op amp device is a little more complex:

$$G = 1 + (R_f/R_i)$$

The equation for a Norton amplifier is obviously simpler to work with. But the easy equation is not the only advantage of using a Norton op amp in a noninverting amplifier circuit.

In a conventional op amp circuit, the minimum gain for a noninverting amplifier is unity (1). This minimum gain can only be achieved by making both resistors, R_f and R_i, equal to zero:

$$\begin{aligned}G &= 1 + R_f/R_i \\ &= 1 + 0/0 \\ &= 1 + 0 \\ &= 1\end{aligned}$$

In this type of circuit, negative gains (attenuation) are impossible.

In the Norton op amp noninverting amplifier circuit however, negative gains are easily achieved, simply by giving the input resistor, R_i, a larger value than the feedback resistor R_f. Unity gain is obtained whenever these two resistors, R_i and R_f, have equal values.

Typical Applications

Of course differential circuits, using both inputs are also possible with a Norton op amp.

The Norton op amp also has many other uses, many of which will be dealt with in the projects of chapters 9 through 11.

Some popular uses for Norton op amps include:

- Voltage comparators
- Voltage regulators
- Rectangular-wave generators
- VCOs
- Precision current sources
- Precision current sinks

and many others.

But before we get to the actual projects, let's take a look at a practical Norton operational amplifier device: the LM3900.

❖ 8
The LM3900

WITHOUT A DOUBT, THE MOST FAMILAR AND WIDELY AVAILABLE Norton op amp device around is the LM3900. In fact, this may well be the only chip of this type the average electronics hobbyist ever encounters. The LM3900 contains four separate and independent Norton amplifiers in a single 14-pin DIP housing.

This IC was first developed in the early 1970s and it is still in widespread use. This is a strong testament to the versatility of the LM3900. The original design goals for this chip included low cost, good, if not high-precision performance, wide output voltage swing, and operation from a single-ended power supply.

SPECIFICATIONS

The specifications for the LM3900 quad Norton amplifier IC are fairly impressive.

Unlike most conventional op amp chips, this device can be operated from a single-polarity supply voltage. The IC's voltage requirements permit an impressively wide range of possible supply voltages. The LM3900 can be powered by anything from a low of +4 volts up to a maximum of +36 volts.

If it is more appropriate to your need, a dual-polarity power supply may be used with the LM3900. In this case, the negative supply voltage, V−, is applied to the ground pin (pin 7). No direct ground connection is made to the chip. If a dual-polarity power supply is used to operate the LM3900, the supply voltages may range from ±2 volts up to ±18 volts. The current drain the chip draws from the power supply is independent of the supply voltage.

A single set of power supply connection pins is included on the IC. These supply voltage pins serve all four of the Norton op amps on the chip. All other functions of the four Norton amplifiers are fully independent and brought out to their own individual pins.

Each Norton amplifier stage has three dedicated pin connections: two inputs, inverting and noninverting, and one output. Additional connection points found on some conventional op amp devices, to allow for special biasing or frequency compensation, are not needed with the LM3900.

Each of the four Norton amplifier stages in the LM3900 is capable of an open loop gain of 70 dB. When one of the Norton op amps is set for unity gain, its bandwidth will be about 2.5 MHz. The bandwidth varies with gain, so this specification is accurate only at unity gain.

The input currents applied to an LM3900 Norton amplifier should be relatively small. For most practical uses, the range of acceptable input currents runs from about 0.5 μA up to approximately 500 μA. As a general rule, the ideal input bias current for this device is about 10 μA.

The output voltage from each of the LM3900's Norton amplifiers can swing over a fairly wide range. The peak-to-peak voltage swing is approximately 1 volt less than the full supply voltage applied to the circuit. In other words, if the LM3900 is being run from a +15-volt power supply, the output voltage can vary over about a 14-volt range.

Each of the Norton amplifier's outputs is short circuit protected. Each of the four Norton amplifiers contained within the LM3900 is internally frequency compensated. No external frequency compensation components are required in circuits built around this device.

The required input biasing current is so low, the value is practically negligible. The stated value in the manufacturer's specification sheet is just 30 nanoamperes (nA). One nanoampere is one billionth of an ampere.

INTERNAL STRUCTURE

The pin-out diagram for the LM3900 quad Norton op amp IC is shown in Fig. 8-1. This chip is housed in a standard 14-pin DIP package.

Internal Structure   137

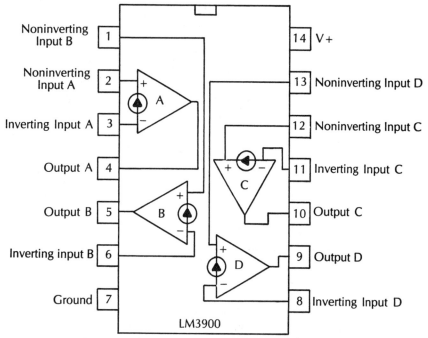

Fig. 8-1 *The LM3900 contains four independent Norton amplifier stages in a single 14-pin chip.*

Aside from the power supply connections, all of the remaining twelve pins are either inputs or outputs, implying that this device is quite easy to use in a wide variety of circuits.

A simplified circuit diagram of one of the Norton amplifier sections is shown in Fig. 8-2. This circuit is repeated a total of four times on the chip. The only common connections between these four Norton amplifier stages are V+, pin 14, and ground, pin 7.

This circuit is a *current-differencing amplifier*. A conventional op amp, on the other hand, is a *voltage differencing amplifier* (VDA). It generally isn't very easy to make a VDA function as a CDA, but it is quite simple to make a CDA simulate the operation of a VDA. All you have to do is place a pair of high value resistors in series with the CDA's inputs. Following Ohm's Law:

$$I = E/R$$

138   The LM3900

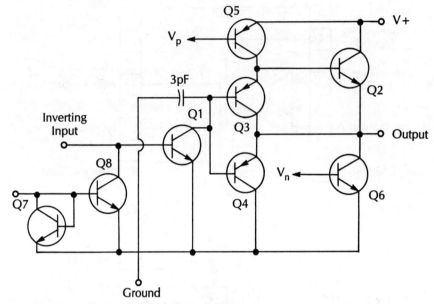

Fig. 8-2 *This is a simplified circuit diagram of one of the Norton amplifier in the LM3900 IC.*

In this case, the resistance, R, has a constant value. Therefore, the current, I, varies in direct proportion to the voltage, E.

The input bias current applied to the base of transistor Q1 is normally equal to the difference between the two input (inverting and noninverting) currents. As stated earlier, the preferred range for these currents in most uses is about 0.5 µA to 500 µA.

If the current at the base of Q1 (input bias current) is higher than about 30 nA, the Norton amplifier's output voltage will start to drop to the nominal zero input voltage, or one-half the circuit's supply voltage.

The output of each of the four Norton amplifiers on the LM3900 is protected against short circuits, increasing the overall reliability of the chip.

BREAKING DOWN THE CIRCUITRY

To better understand the internal circuitry of the LM3900 Norton operational amplifier IC, we will break it down into simpler subcircuits. The full circuit is illustrated in Fig. 8-2. The transistor numbering from this diagram will be maintained in the diagrams for the subcircuits to allow you to more conveniently compare the schematics.

Figure 8-3 shows a simplified version of the very basic inverting amplifier circuit. Notice that we are looking at only transistors Q1 and Q2 here. For the time being, we will ignore the rest of the circuitry of Fig. 8-2.

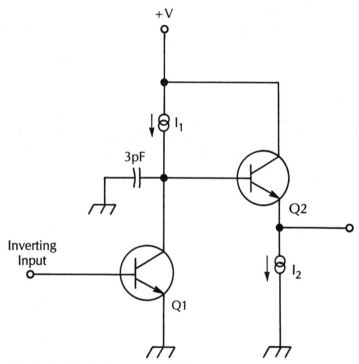

Fig. 8-3 *This is the LM3900's basic inverting amplifier circuit.*

Transistor Q1 is wired as a simple common-emitter amplifier. This type of circuit inverts the signal polarity. That is, a positive input results in a negative output and vice versa.

The collector load for this one transistor amplifier maintains a constant current of 200 μA. Transistor Q1 is followed by transistor Q2 which is a noninverting emitter follower, serving as a buffer. The emitter load of this transistor is also a constant current, this time with a value of 1300 μA.

This amplifier stage has a very high gain. The upper frequency response is limited by the 3 pF capacitance. This is an intentional trade-off in the circuit design. This frequency limiting capacitance is included to improve the stability of the circuit. Without this capacitance, the amplifier circuit might be prone to uncontrollable oscillations under certain conditions.

Transistors Q1 and Q2 combine to form a high-gain inverting amplifier stage. The maximum overall current gain of this subcircuit is equal to the product of the current gain for each of the individual transistors ($Q_1 \times Q_2$).

The output of this subcircuit is a voltage which can swing essentially over the entire supply voltage range. The minimum output voltage is a few hundred millivolts above ground, and the maximum output voltage is a few hundred millivolts less than the actual supply voltage applied to the circuit.

Adding transistor Q3 to the circuit, as illustrated in Fig. 8-4 significantly increases the overall current gain, with only a minimal decrease in the output voltage range. Notice that transistors

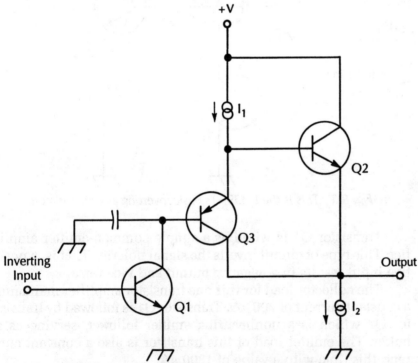

Fig. 8-4 *The addition of transistor Q3 greatly increases the gain of the inverting amplifier circuit of Fig. 8-3.*

Q2 and Q3 are complementary types. Q2 is an npn transistor, and Q3 is a pnp transistor.

This modified subcircuit can put out (source) up to 10 mA. However, the constant current source in Q2's emitter line limits its current sinking capabilities to 1.3 mA.

A more complete version of this subcircuit is shown in Fig. 8-5. Transistors Q5 and Q6 are constant current generators, which are internally biased within the LM3900. Transistor Q4 operates as a class-B amplifier when the circuit is in an overdrive condition. This greatly increases the potential sink current of the circuit.

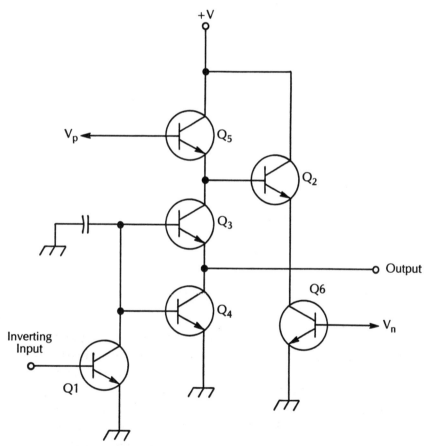

Fig. 8-5 *This is a more complete version of the LM3900's basic inverting amplifier circuit.*

The circuit of Fig. 8-5 is the heart of each of the Norton amplifier stages within the LM3900. However, so far we do not have an operational amplifier, because this subcircuit is capable of only inverting operation. Noninverting and differential circuits are not yet allowed for.

A noninverting input is added to the basic inverting amplifier circuit with transistors Q7 and Q8, as illustrated in Fig. 8-6. These two transistors make up a current mirror subcircuit. The two transistors are closely matched, with identical operating characteristics. This subcircuit will draw an output current which is, for all practical purposes, equal to the input current.

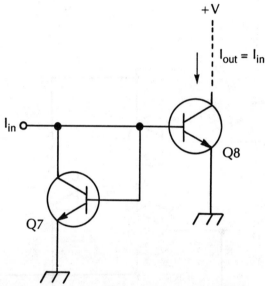

Fig. 8-6 A current mirror is used to add a noninverting input to the basic inverting amplifier circuit of Fig. 8-5.

The current mirror's input current is applied to the bases of both transistors Q7 and Q8. For the next part of our discussion, we will have to make a few assumptions to give us some definite values to work with. Let's assume that the gain for each transistor is 100. Remember Q7 and Q8 are identical units, so their gain must be the same. We will also assume that the base current drawn by each transistor is 6 μA. The collector current of each transistor is simply the product of the base current and the transistor's gain, or 600 μA in our example.

Notice that the collector current for transistor Q7 is drawn directly from the circuit's input current. This changes the current value at this point in the circuit. The input current must provide the collector current of transistor Q7 and the base current of both transistors. In algebraic terms:

$$I_{c7} = (I_b \times G_7) + I_b + I_b$$
$$= (I_b \times G_7) + 2I_b$$

where I_b is the base current drawn by either transistor, G_7 is the gain of transistor Q_7, and I_{c7} is the total current at the collector of this transistor.

For example, this works out to:

$$I_{c7} = (6 \times 100) + (2 \times 6)$$
$$= 600 + 12$$
$$= 612 \; \mu A$$
$$= 0.000612 \; \text{ampere}$$

Meanwhile, the collector current of transistor Q8 is the output current of this subcircuit. The output current of this subcircuit may also be called the *mirror current*. The input and output currents will always be very nearly equal, regardless of the magnitude of the actual input current.

The circuit shown in Fig. 8-2 is simply the combination of the subcircuits shown in Figs. 8-5 and 8-6. The Norton amplifier's inverting input obviously feeds the inverting amplifier subcircuit of Fig. 8-5, and the noninverting input feeds the input of the current mirror subcircuit of Fig. 8-6. The mirror (output) current is drawn from the chip's inverting input. As a result of these connections, the base current of transistor Q1 is always equal to the inverting input current minus the noninverting input current.

In effect, the combination of the inverting amplifier subcircuit of Fig. 8-5 and the current mirror subcircuit of Fig. 8-6 form a practical CDA circuit.

The circuitry has been somewhat simplified here for the purposes of discussion. Of course, it is important to recall that this CDA circuit is repeated four separate times within the LM3900 IC.

❖ 9
Amplifier Projects

THE FINAL THREE CHAPTERS OF THIS BOOK WILL PRESENT ABOUT two dozen practical projects using the LM3900 quad Norton operational amplifier IC. The projects presented in this chapter cover various basic amplification uses.

INVERTING AMPLIFIER

A practical inverting amplifier circuit using a Norton op amp is illustrated in Fig. 9-1. A typical parts list for this project appears in Table 9-1. Feel free to experiment with different component values.

This circuit is intended for use with ac signals. Capacitor C1 blocks any dc component in the input signal. If you want to use the circuit as a dc amplifier, simply omit the input capacitor. If the capacitor is used, its value is not particularly critical.

Resistor R1 converts the input voltage into an input current acceptable by the Norton amplifier's inverting input.

The noninverting input is shorted to the circuit's positive supply voltage through resistor R3. The values of resistors R2 and R3 are selected to set the quiescent output voltage. This is the output voltage which will be present when the input signal equals zero, or if the input is grounded.

For most linear applications, the quiescent output voltage is set at the midpoint of the supply voltage. That is, if the circuit is being operated off of a +15-volt power supply, the quiescent output voltage is set at one-half this value, or +7.5 volts. This gives the output voltage an equal swing range on either side of the quiescent point. In other words, the input and output signals can go either positive or negative.

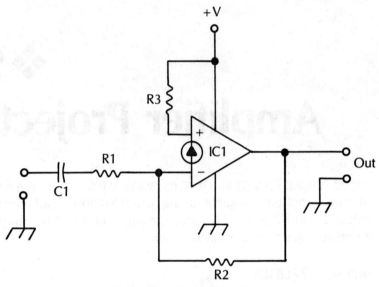

Fig. 9-1 Inverting Amplifier

Table 9-1 Parts List for the Inverting Amplifier project of Fig. 9-1.

Part	Description
IC1	LM3900 quad Norton op amp
C1	0.1 µF capacitor
R1	100K 1/4-watt 5 percent resistor
R2	470K 1/4-watt 5 percent resistor
R3	1 Megohm 1/4-watt 5 percent resistor

To maintain the desired V+/2 quiescent output voltage, the bias resistor, R3, should have a value about twice that of the feedback resistor (R2).

As explained in chapter 7, the amplifier gain is determined by the values of the input resistor, R1, and the feedback resistor, R2. The formula is simply:

$$G = -R_2/R_1$$

The negative sign indicates that the signal polarity is inverted at the output. The output signal is 180 degrees out of phase with the input signal. If the input voltage is positive, the output voltage will be negative, and vice versa.

As a demonstration of the gain equation, let's use the component values suggested in the parts list (Table 9-1):

$$R_1 = 100\text{K}\Omega$$
$$R_2 = 470\text{K}\Omega$$

In this case the circuit's gain works out to:

$$G = -470,000/100,000$$
$$= -4.7$$

or approximately −5.

Because feedback resistor R2 interacts with resistor R3 to set the quiescent output voltage, it is best to change the amplifier gain by selecting a new value for the input resistor, R1. Increasing the value of this resistor will decrease the circuit gain. For example, if we boost the value of resistor R1 to 220KΩ, the gain of the amplifier drops to about:

$$G = -470,000/220,000$$
$$= -2.1$$

Conversely, the gain of the amplifier can be increased by giving resistor R1 a smaller value. As an example, let's drop the value of resistor R1 to 68K. This change increases the circuit's gain to approximately:

$$G = -470,000/68,000$$
$$= -6.9$$

If the value of input resistor R1 is made larger than the value of the feedback resistor, the gain becomes negative, or less than unity. In other words, the signal is attenuated rather than amplified. The signal amplitude is less at the output of the circuit than at the input.

For instance, if we increase the value of resistor R1 to 1 megohm, the gain works out to a value of:

$$G = -470,000/1,000,000$$
$$= -0.47$$

or a little under −0.5. The output signal has an amplitude about half of the input signal's amplitude.

You are strongly encouraged to experiment with alternative component values in this circuit. Try various gains. Try different combinations of values for resistors R2 and R3. How does the circuit's operation change if the quiescent output voltage is not set at half the supply voltage?

NONINVERTING AMPLIFIER

The circuit for a noninverting amplifier built around a Norton op amp is very similar to the inverting amplifier circuit presented in the last section. The noninverting version is shown in Fig. 9-2. A suitable parts list for this project is given in Table 9-2. Experiment with different component values.

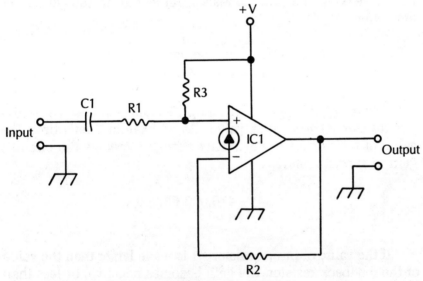

Fig. 9-2 *Noninverting Amplifier*

The only real difference between this noninverting circuit and the inverting circuit of Fig. 9-1 is that in this project the input signal is applied through capacitor C1 and resistor R1 to the Norton op amp's noninverting input, instead of to its inverting input. As a result, the output signal will be in phase with the input signal. There is no polarity inversion in this circuit.

Table 9-2 Parts List for the Noninverting Amplifier project of Fig. 9-2.

Part	Description
IC1	LM3900 quad Norton op amp
C1	0.1 μF capacitor
R1	100K ¼-watt 5 percent resistor
R2	1 Megohm ¼-watt 5 percent resistor
R3	2.2 Megohm ¼-watt 5 percent resistor

As with the preceding project, this circuit is intended for use with ac signals. Capacitor C1 blocks any dc component in the input signal. If you want to use the circuit as a dc amplifier, simply omit the input capacitor. If the capacitor is used, its value is not critical.

Resistor R1 converts the input voltage into an input current acceptable by the Norton amplifier's noninverting input.

Feedback resistor R2 connects the inverting input to the Norton amplifier's output. The negative feedback limits the open loop gain of the device.

The values of resistors R2 and R3 are selected to set the quiescent output voltage. This is the output voltage which will be present when the input signal equals zero, or if the input is grounded.

For most linear applications, the quiescent output voltage is set at the mid-point of the supply voltage. That is, if the circuit is being operated from a +12 volt power supply, the quiescent output voltage is set at one-half this value, or +6 volts. This gives the output voltage an equal swing range on either side of the quiescent point. To accomplish this, bias resistor R3 should have a value that is about twice that of feedback resistor R2.

Determining the amplifier's gain is quite simple. The gain in this circuit is set by the values of input resistor R1 and feedback resistor R2. The gain formula for a noninverting Norton amplifier circuit is simply:

$$G = R_2/R_1$$

Notice that there is no negative sign in this equation, indicating that there is no polarity inversion of the input signal at the output of this circuit.

150 Amplifier Projects

As a demonstration of the gain equation, let's use the component values suggested in the parts list (Table 9-2):

$$R_1 = 100\text{K}\Omega$$
$$R_2 = 1 \text{ Megohm}$$

In this case the circuit's gain works out to:

$$G = 1{,}000{,}000/100{,}000$$
$$= 10$$

Because feedback resistor R2 interacts with resistor R3 to set the quiescent output voltage, it is usually more convenient to change the amplifier gain by selecting a new value for input resistor R1. Increasing the value of this resistor will decrease the circuit gain. For example, if we boost the value of resistor R1 to 220K, the gain of the amplifier drops:

$$G = 1{,}000{,}000/220{,}000$$
$$= 4.5$$

Conversely, the gain of the amplifier can be increased by giving resistor R1 a smaller value. As an example, let's drop the value of resistor R1 to 68K. This change increases the circuit's gain to approximately:

$$G = 1{,}000{,}000/68{,}000$$
$$= 14.7$$

If the value of R1 is made larger than the value of the feedback resistor, the gain becomes negative, or less than unity. In other words, the signal is attenuated rather than amplified. The signal amplitude is less at the output of the circuit than at the input.

For instance, if we increase the value of resistor R1 to 2.2 megohms, the gain works out to a value of:

$$G = 1{,}000{,}000/2{,}200{,}000$$
$$= 0.45$$

or a little under 0.5. The output signal has an amplitude of about half of the input signal's amplitude.

As with the inverting amplifier project presented earlier in this chapter, you are strongly encouraged to experiment with alternative component values in this circuit. Try various gains. Try different combinations of values of resistors R2 and R3. How does the circuit's operation change if the quiescent output voltage is not set at half the supply voltage?

DIFFERENTIAL AMPLIFIER

Unlike either an inverting amplifier or a noninverting amplifier, a differential amplifier uses both of a Norton op amp's inputs simultaneously, as shown in the circuit diagram of Fig. 9-3.

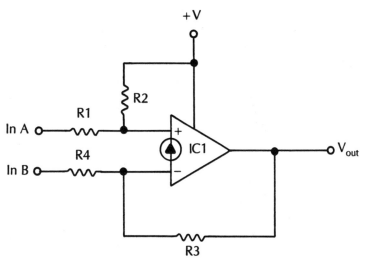

Fig. 9-3 Differential Amplifier

A typical parts list for this project appears as Table 9-3. Once again, you are encouraged to experiment with other component values.

Table 9-3 Parts List for the
Differential Amplifier project of Fig. 9-3.

Part	Description
IC1	LM3900 quad Norton op amp
R1, R4	100K 1/4-watt 5 percent resistor
R2	470K 1/4-watt 5 percent resistor
R3	1 Megohm 1/4-watt 5 percent resistor

Most differential amplifier applications include dc, or very slowly changing signals, so no input capacitors are shown in this circuit. If appropriate to your specific use, it is a simple enough matter to add a dc blocking capacitor before input resistors R1 and R4. A good value for such a capacitor is in the 0.05 to 0.5 μF range.

Both input resistors, R1 and R4, should have equal values so the two input signals are similarly weighted. In most applications, both the inverting input signal and the noninverting input signal should be subjected to an identical amount of gain.

As with the inverting amplifier and noninverting amplifier circuits presented earlier in this chapter, when the difference between the inputs is exactly zero, the quiescent output voltage should generally be set to one-half of the supply voltage. This is done by giving bias resistor R3 a value equal to about twice the value of feedback resistor R2.

The gain for the inverting and the noninverting signals is controlled separately. For the signal applied to the inverting input, the gain is equal to:

$$G_i = -R_2/R_1$$

The negative sign in this equation indicates that the signal polarity is inverted for this input.

The gain equation for the noninverting input is very similar, except there is no polarity inversion, and a different input resistor is used:

$$G_n = R_2/R_4$$

The output voltage (V_{oa}) resulting from the signal at the inverting input is equal to the inverting input voltage (V_{ia}) multiplied by the inverting gain (G_i). Remember, the input resistor converts the input voltage into an input current. In algebraic terms:

$$-V_{oa} = V_{ia} \times G_i$$

Remember this gain (G_i) is always a negative number, indicating the polarity inversion of the Norton amplifier's inverting input. Therefore, V_{oa} is always at the opposite polarity as V_{ia}.

Differential Amplifier

The output voltage (V_{ob}) due to the noninverting signal is calculated similarly. Multiply the voltage at the noninverting input (V_{ib}) by the noninverting gain (G_n):

$$V_{ob} = V_{ib} \times G_n$$

This gain value, G_n, is always positive because the polarity is not inverted.

The total output voltage from the differential amplifier is simply equal to the difference between the inverting output voltage (V_{oa}) and the noninverting output voltage (V_{ob}):

$$V_o = V_{ob} - V_{oa}$$

In virtually all differential amplifier circuits, the same gain will be used for both of the input signals. This is done by giving input resistors R1 and R4 equal values:

$$R = R_1 = R_4$$

This makes the gains equal, except for their polarity:

$$G_i = -R_2/R$$
$$G_n = R_2/R$$

The output voltage of the circuit then becomes equal to:

$$V_o = (V_{ib} - V_{ia}) \times G$$

In the suggested parts list of Table 9-3, equal values are used for input resistors R1 and R4, and feedback resistor R2. This gives us unity gain, simplifying the output voltage equation to:

$$V_o = (V_{ib} - V_{ia}) \times 1$$
$$= V_{ib} - V_{ia}$$

The output voltage is simply equal to the difference between the two input voltages. The voltage applied to the inverting input, through input resistor R1, is subtracted from the voltage applied to the noninverting input through input resistor R4.

154 Amplifier Projects

The operation of this circuit is made clear with a few practical examples:

Inverting Input	Noninverting Input	Output
0	0	0 volt
1	0	−1 volt
0	1	1 volt
1	1	0 volt
2	1.5	−0.5 volt
3.5	4	0.5 volt
2	3	1 volt
4	3	−1 volt

We can increase the gain by reducing the value of input resistors R1 and R4. For example, we will change the value of both resistor R1 and resistor R4 to 470K. Both input resistors still have equal values. The gain equation for the circuit now gives us a value of:

$$G = R_2/R$$
$$= 1{,}000{,}000/470{,}000$$
$$= 2.13$$

We can round this off to a gain of two. The tolerances of practical resistors are likely to result in at least that much error. Now, the output voltage is equal to the difference of the two input voltages multiplied by two:

$$V_o = (Vi_b - Vi_a) \times 2$$

Repeating our examples for the modified circuit, we get the following results:

Inverting Input	Noninverting Input	Output
0	0	0 volt
1	0	−2 volts
0	1	2 volts
1	1	0 volt
2	1.5	−1 volt
3.5	4	1 volt
2	3	2 volts
4	3	−2 volts

In most cases, a differential amplifier's inputs are dc voltages. However, some very interesting results can be obtained by experimenting with ac signals to one or both of the inputs of a differential amplifier circuit.

WIDE-BANDWIDTH/HIGH-GAIN AMPLIFIER

An ideal amplifier circuit would have an infinite bandwidth. Any input signal, regardless of its frequency, would receive exactly the same amount of amplification with no distortion.

In practical circuits, there are always limitations and trade-offs. No real amplifier circuit can handle all possible frequencies equally. As a somewhat extreme example, there are usually several significant differences between an amplifier circuit designed for audio frequencies (AF) compared to a radio frequency (RF) amplifier circuit.

Simple amplifier circuits generally can't even handle the audible frequency spectrum, approximately 20 Hz to 20,000 Hz, with a true flat frequency response. For most practical uses, the wider the bandwidth, the better the amplifier circuit.

The LM3900 Norton op amp's frequency response is internally limited. This is indicated by the slew rate specification. The slew rate, you should recall, is a measurement of how fast the output signal can change, in response to a rapid change in the input signal. The Norton amplifiers in the LM3900 chip are rated for a slew rate of 0.5 volt per second. This limits the high frequency response of the device.

It is very important to realize that this frequency limitation is not due to poor design. The slew rate of the LM3900 Norton op amp was purposely limited to improve the stability of the IC. Practical circuit design very often requires these trade-offs. In an amplifier circuit, stability must be balanced against bandwidth. If the bandwidth is made too wide, the amplifier may become unstable and break into uncontrollable oscillations under certain operating conditions, especially when high gains are involved.

The LM3900 Norton amplifier was designed for high stability without the necessity of any external frequency compensation components. This simplifies circuit design using this IC for most applications.

In some practical uses, however, we might want to get around the internal limitations of the device. This can often be done with a few extra components.

The circuit illustrated in Fig. 9-4 is a wide-bandwidth/high-gain amplifier. A suitable parts list for this project is given in Table 9-4.

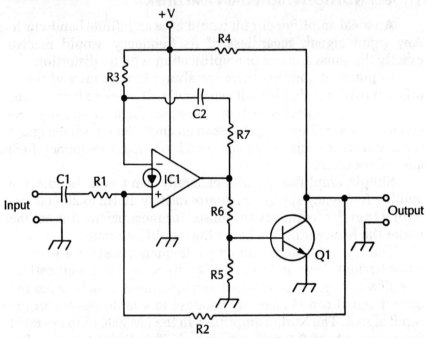

Fig. 9-4 Wide-bandwidth/High-gain Amplifier

**Table 9-4 Parts List for the
Wide-bandwidth/High-gain Amplifier
project of Fig. 9-4.**

Part	Description
Q1	NPN 2N2222 or 2N3904
IC1	LM3900 quad Norton op amp
C1	0.1 µF capacitor
C2	25 pF capacitor
R1, R4	10K 1/4-watt 5 percent resistor
R2, R7	1 Megohm 1/4-watt 5 percent resistor
R3	2.2 Megohm 1/4-watt 5 percent resistor
R5, R6	2.7K 1/4-watt 5 percent resistor

Transistor Q1 may be almost any low power npn type designed for audio frequency use. The 2N2222, and the 2N3904 are widely available and inexpensive.

This circuit is designed as a noninverting amplifier. If you need an inverting amplifier, it is a very easy modification to make. Just move input resistor R1 from the IC's inverting input to the noninverting input. That's the only change required.

The input capacitor, C1, blocks any dc component in the input signal. If you need extremely low frequency response, or amplification of dc signals, you may omit this capacitor from the circuit.

Using the component values from the suggested parts list, this amplifier circuit has a gain of about 100, and can pass frequencies up to nearly 200 kHz. This is well beyond the upper limit of the audible range, which only extends up to about 20 kHz, or less.

The frequency response of the circuit won't be completely flat up to the maximum frequency limit (200 kHz). There will be some fluctuations here and there along the way, especially at higher frequencies. But by making the bandwidth so wide, the frequency response will be extremely flat in the audible range.

The overall frequency response of the circuit can be made a little flatter in some special applications by deleting capacitor C2 and resistor R7 from the circuit. Generally, however, it is best to include these two components.

The function of capacitor C2 and resistor R7 is to improve the stability of the circuit. This amplifier has quite a large gain. Under certain conditions, the circuit may be prone to function as an oscillator instead of an amplifier if stabilizing components C2 and R7 are not included in the circuit.

Another advantage of this particular circuit is that the output voltage can swing over a larger than usual range. It is not limited to the supply voltage.

HIGH-PASS FILTER

It usually isn't worded in such terms, but a *filter* could be defined as "an amplifier circuit with a deliberately restricted bandwidth." A filter circuit passes certain frequency components, while blocking, or significantly attenuating others.

A filter circuit is useful when a signal must be simplified (the number of frequency components reduced), or when certain signals, such as a 60 Hz hum, must be blocked from certain portions of an electronic system.

The difference between a filter circuit and a badly designed amplifier circuit is that in a filter, the limited frequency response is very specific and predictable. Although there are others, the four basic types of filter circuits are:

- low-pass filters
- high-pass filters
- band-pass filters
- band-reject filters

These names are all self-explanatory. Also, refer back to the VCF projects in chapter 6.

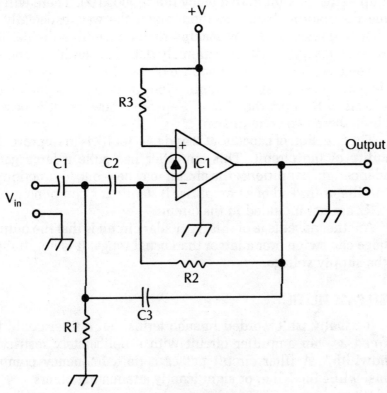

Fig. 9-5 *High-pass Filter*

High-pass Filter 159

Table 9-5 Parts List for the High-pass Filter project of Fig. 9-5.

Part	Description
IC1	LM 3900 quad Norton op amp
C1, C2, C3	0.001 µF capacitor
R1	4.7K ¼-watt 5 percent resistor (see text)
R2	470K ¼-watt 5 percent resistor
R3	1 Megohm ¼-watt 5 percent resistor

A high-pass filter circuit built around one section of an LM3900 Norton amplifier IC is illustrated in Fig. 9-5. A suitable parts list for this project is given in Table 9-5. This is a good project for additional experimentation. Breadboard the circuit, and try some alternative component values.

A high-pass filter is designed to permit any frequency component higher than the cutoff frequency to reach the output unhindered. All frequencies below the cutoff frequency are blocked, or significantly attenuated, by the filter circuit. A simple frequency response graph for a high-pass filter like this project is shown in Fig. 9-6.

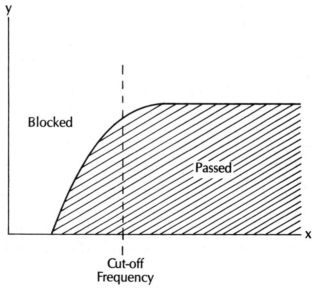

Fig. 9-6 A high-pass filter blocks low frequencies, while passing higher frequencies on to the output.

The cutoff frequency is defined as the frequency point where the signal is attenuated by a factor of 6 dB. Notice how the input signal must pass through capacitors C1 and C2, and the feedback signal must pass through capacitors C3 and C2. Remember, a capacitor blocks dc voltages, and has greater impedance at low frequencies than at high frequencies.

For this circuit to function properly, all three capacitors should have the same value. For use with audio frequency signals, these capacitors will have relatively low values.

As with all the other Norton amplifier projects in this chapter, the quiescent output voltage is biased to half the supply voltage by making resistor R3 twice as large as resistor R2.

Virtually all the passive components in this circuit contribute to determining the filter's cutoff frequency. The approximate cutoff frequency can be calculated as:

$$F = \frac{(\sqrt{R_2/R_1} \times 2) - 3.5}{6.28 \times C \times R_2}$$

where R_1 and R_2 are the input and feedback resistances in ohms, C is the capacitance value in farads, used for all three capacitors in the circuit, and F is the cutoff frequency, in Hertz.

Using the component values suggested in the parts list, the filter's cutoff frequency works out to about 3 kHz. It is easiest to change the cutoff frequency by substituting a different input resistor value for R1. To change C would require changing all three capacitors. Altering the value of the feedback resistor, R2, would require also changing bias resistor R3. Input resistor R1 can be changed without altering anything else in the circuit.

Increasing the value of input resistor R1 decreases the filter's cutoff frequency. Also, the cutoff frequency can be raised by giving input resistor R1 a smaller value.

Here are some additional examples using several common resistor values for input resistor R1. All other component values in the circuit (R2, R3, C1, C2, and C3) are assumed to be the same as those specified in the parts list of Table 9-5. The resulting cutoff frequency in each case is rounded off to the nearest kiloHertz:

R1	F
1K	7000 Hz
1.5K	5000 Hz

R1	F
2.2K	4000 Hz
4.7K	3000 Hz
10K	2000 Hz
22K	1000 Hz

Notice that the cutoff frequency change is not linear with respect to the input resistance R1. The cutoff frequency for this high-pass filter circuit is fixed by the component values used. There is no provision for a voltage control input in this circuit, unlike the VCF projects presented in chapter 6.

❖ 10
Voltage and Current Projects

IN THE PROJECTS OF THIS CHAPTER, THE NORTON AMPLIFIER IS used to directly manipulate and control voltages and currents in various ways. Most of these are not stand-alone projects. These circuits are intended for use with other electronic circuits, as part of a larger system.

VOLTAGE REGULATOR

More and more electronics circuits these days require a regulated supply voltage. This is especially true of digital circuitry. A regulated voltage has very little fluctuation, or *ripple*. There is virtually no ac component in a regulated dc voltage. A regulated voltage is also one which does not vary with changes in the load.

A practical voltage regulator circuit built around an LM3900 Norton op amp is illustrated in Fig. 10-1. A suitable parts list for this project is given in Table 10-1.

The zener diode, D1, and npn transistor Q1 should be selected to suit the intended use. The zener diode's breakover voltage should be equal to the desired output (regulated) voltage.

The transistor must be selected to handle the maximum power that will ever be drawn from the voltage regulator circuit. The current rating of this device is very important here. A good rule is to choose a transistor which is rated for at least 20 to 30 percent more current than your intended load circuit will ever draw from the voltage regulator. When in doubt, use a heftier transistor.

This circuit is fairly simple and inexpensive, but it is quite effective, and should provide adequate voltage regulation for all

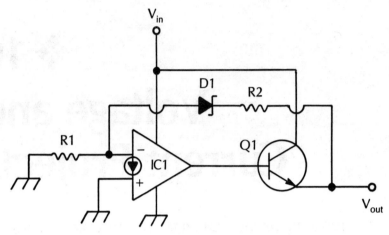

Fig. 10-1 Voltage Regulator

Table 10-1 Parts List for the Voltage Regulator project of Fig. 10-1.

Part	Description
IC1	LM3900 quad Norton amplifier IC
Q1	NPN transistor (2N3904, or similar)
D1	zener diode—see text
R1	1K 5 percent 1/4-watt resistor
R2	100 ohm 5 percent 1/4-watt resistor

but the most critical applications. Notice that the input voltage to this circuit is also the supply voltage for the Norton amplifier (IC1).

The unregulated input voltage must be at least two volts higher than the desired regulated output voltage. The voltage regulator itself necessarily consumes a certain amount of power, and causes a voltage drop. The regulation of the output voltage is likely to suffer if the input voltage is too low.

The input voltage should be rectified. It would also be a good idea to use some filtering on the input voltage before applying it to this circuit. Don't expect any practical voltage regulator to do all the work. The voltage regulator will do its job best if the input voltage is already reasonably close to a constant dc voltage.

VARIABLE VOLTAGE REGULATOR

The voltage regulator circuit shown in Fig. 10-2 is even more versatile than the preceding project. The regulated output voltage from this circuit is manually adjustable. A suitable parts list for this project appears as Table 10-2.

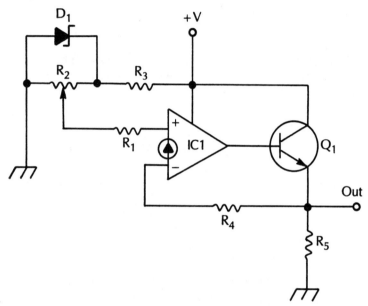

Fig. 10-2 Variable Voltage Regulator

Table 10-2 Parts List for the Variable Voltage Regulator project of Fig. 10-2.

Part	Description
IC1	LM 3900 quad Norton amplifier IC
Q1	NPN power transistor (2N3055, or similar)
D1	15 Volt zener diode
R1	470K 5 percent 1/4-watt resistor
R2	10K potentiometer
R3	2.7K 5 percent 1/4-watt resistor
R4	1 Megohm 5 percent 1/4-watt resistor
R5	10K 5 percent 1/4-watt resistor

Basically, in this circuit, the Norton amplifier, IC1, is being used as a noninverting dc amplifier. The gain of this amplifier is determined by the values of input resistor R1, and feedback resistor R4.

The gain formula for the Norton op amp, as you should recall from the previous chapters, is:

$$G = R_f/R_i = R_4/R_1$$

In this application, the gain is fairly low. The parts list, Table 10-2, recommends the following values for the gain determining resistors, R1 and R4:

$$R_1 = 470 \text{K}\Omega$$
$$R_4 = 1 \text{ Megohm}$$

Using these resistor values, the amplifier's gain works out to:

$$G = 1000000/470000 = 2.13$$

or about 2.

The input voltage to this dc amplifier is set up by a voltage divider network made up of zener diode D1, resistor R3, and potentiometer R2. Adjusting the potentiometer alters the input voltage, which obviously results in a change in the circuit's output voltage.

The value of the zener diode essentially controls the range of the variable voltage regulator's output. The output voltage will never exceed a value equal to approximately twice the zener diode's breakover voltage.

A 15-volt zener diode is suggested for D1 in the parts list. This may be reduced if you need a narrower range of possible output voltages. Increasing the voltage of the zener diode is not recommended. Too high an input voltage will force the Norton amplifier into output clipping, and might conceivably cause damage to the IC, or possibly to other components in the regulator circuit or the load.

The minimum output voltage is determined primarily by the maximum resistance of potentiometer R2, and by the value of resistor R3. Using the component values recommended in the parts list, the minimum output voltage from this circuit is about 0.5 volt.

The full range of this variable voltage regulator project, assuming all of the component values from the parts list are used in the circuit, runs from about 0.5 volt to approximately 30 volts. This range should cover most practical needs in hobbyist applications.

Transistor Q1 boosts the output current. It must be an npn type; otherwise, the choice of the transistor is not too critical, as long as it can supply more than the maximum current drawn by your intended load. Overrate the current handling capability of the output transistor by at least 20 to 30 percent.

The 2N3055 NPN power transistor suggested in the parts list is a fairly hefty device. It is relatively inexpensive and easy to find on the hobbyist market. The 2N3055 should be able to supply adequate current for the majority of practical uses.

VARIABLE-REFERENCE VOLTAGE SOURCE

Many practical circuits require a reference voltage of some kind. Several of the projects presented earlier in this book have this requirement.

Figure 10-3 shows a simple reference-voltage source circuit. The output voltage can be set with fair precision and reliability.

A suitable parts list for this project is given in Table 10-3. This is certainly one of the less complicated projects in this book.

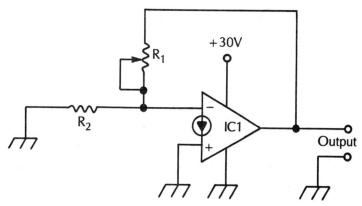

Fig. 10-3 Variable Reference Voltage Source

Table 10-3 Parts List for the Variable Reference Voltage Source project of Fig. 10-3.

Part	Description
IC1	LM3900 quad Norton amplifier IC
R1	500K potentiometer
R2	10K 5 percent ¼-watt resistor

Aside from the LM3900 Norton op amp IC itself, only two external components are required; a potentiometer, R1, and a fixed resistor, R2.

By adjusting the potentiometer over its range, the output voltage from this circuit can be set anywhere from a minimum value of a little over 0.5 volt to a maximum value of about 25 volts. This range of voltage values assumes that the circuit's supply voltage is 30 volts, and the component values suggested in Table 10-3 are used.

In some cases, this circuit could be used as a simple voltage regulator. However, the regulation of the output voltage from this circuit is nowhere as good as that of the circuits shown in Figs. 10-1 and 10-2. The output voltage is likely to fluctuate with changes in ambient temperature.

This circuit is really nothing more than a simple inverting dc amplifier. Depending on the setting of potentiometer R1, the gain can be varied from 1 to about 50. The output of this circuit can supply currents up to several milliamperes.

If you do not need a variable reference voltage, potentiometer R1 can be replaced with a fixed resistor of an appropriate value. The larger this resistance is, the greater the output voltage will be.

SCHMITT TRIGGER

A Schmitt trigger is a circuit which has a HIGH output when the input voltage exceeds a specific, preset level. If the input voltage drops below a specific, preset level, the output goes LOW. There are no in-between output values: the output of a Schmitt trigger is unambiguous.

Schmitt Trigger 169

In general terms, a Schmitt trigger is a specialized type of switching circuit that is frequently used to clean up noisy signals after transmission, storage, or retrieval. A Schmitt trigger can also be used to convert any analog waveshape into a pulse form usable by digital circuitry.

If the input voltage to a Schmitt trigger circuit is very close to the switchover point, the circuit may exhibit "output chatter" in response to random noise in the input signal. A brief burst of noise in the input signal line could cause the circuit to falsely trigger (switch output states). This problem is illustrated in Fig. 10-4. The output may oscillate rapidly back and forth between the HIGH and LOW states under these conditions. In effect, the circuit gets "confused" by the noise.

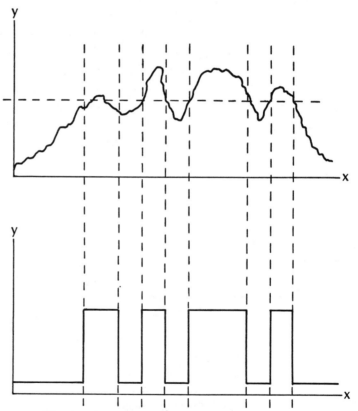

Fig. 10-4 *A noisy input signal near the crossover point can cause a Schmitt trigger circuit to "chatter."*

The solution to such problems is to add some *hysteresis* to the Schmitt trigger circuit. Hysteresis is an important specification for any Schmitt trigger circuit. In very simple terms, hysteresis is a sort of delay in the circuit's response. It might be considered a form of intentional sluggishness.

Common sense would seem to indicate that the less hysteresis, the better. But in many practical uses, the conclusions of common sense don't hold up.

As stated earlier, if a noisy input signal is close to the switchover value, the Schmitt trigger will be prone to chatter. By increasing the hysteresis of the circuit, the turn-on switchover voltage will be higher than the switch-off voltage. This leaves a "dead zone" between the two switching levels, which is ignored by the circuit. Minor noise pulses near the switchover voltages have considerably less effect under these circumstances.

There is a definite trade-off involved with hysteresis, however. A low hysteresis circuit will respond faster to smaller changes in the input signal, while a high hysteresis circuit is less sensitive to noise. The ideal amount of hysteresis will depend on the requirements of the specific application intended.

A simple but practical Schmitt trigger circuit built around an LM3900 Norton amplifier is illustrated in Fig. 10-5. A suitable parts list for this simple project is given in Table 10-4.

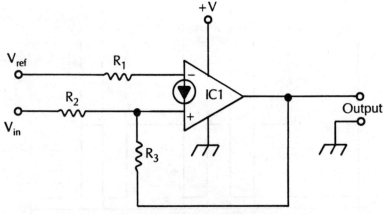

Fig. 10-5 Schmitt Trigger

The hysteresis of this circuit can be adjusted by changing the value of feedback resistor R3. The Schmitt trigger's hysteresis varies with the gain of the amplifier, and thus it is dependent on the

Table 10-4 Parts List for the
Schmitt Trigger project of Fig. 10-3.

Part	Description
IC1	LM3900 quad Norton amplifier IC
R1, R2	1 Megohm 5 percent 1/4-watt resistor
R3	10 Megohm 5 percent 1/4-watt resistor

gain formula:

$$G = R_3/R_2$$

Notice that there are two input signals to this circuit. One input is labeled V_{in}. This is the main input signal to be treated by the Schmitt trigger.

The second input to this circuit is labelled V_{ref}. A constant reference voltage is applied to this input. This would be an appropriate application for the preceding project. The reference voltage is the switchover point of the Schmitt trigger. If input voltage V_{in} exceeds the reference voltage, V_{ref}, the circuit's output goes HIGH. Otherwise, the output of the circuit is LOW.

To change this circuit into an inverting Schmitt trigger is a simple modification. Just reverse the signals applied to the two inputs. Feed the V_{in} signal through resistor R1 to the Norton amplifier's inverting input, and apply the reference voltage, V_{ref}, to the noninverting input through resistor R2.

In virtually all practical uses, both input resistors R1 and R2 should have the same value, which will typically be quite high. If you experiment with this circuit, I'd suggest varying the value of resistor R3, while keeping the values of resistors R1 and R2 at 1 megohm.

A Schmitt trigger circuit like this is basically a modified voltage comparator. For more information on Schmitt triggers, refer to chapter 4.

VOLTAGE COMPARATOR

If we take the Schmitt trigger circuit of Fig. 10-5 and delete feedback resistor R3, we have a simple voltage comparator circuit, as shown in Fig. 10-6. A parts list for this very simple project appears as Table 10-5.

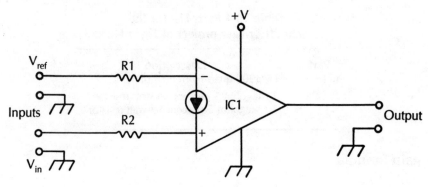

Fig. 10-6 *Voltage Comparator*

Table 10-5 Parts List for the Voltage Comparator project of Fig. 10-6.

Part	Description
IC1	LM3900 quad Norton amplifier IC
R1, R2	1 Megohm 5 percent 1/4-watt resistor

Two input voltages, V_{ref} and V_{in}, are applied to the Norton op amp's two inputs through the input resistors R1 and R2. If the V_{in} voltage is higher than the V_{ref} voltage, the output will be HIGH. This is indicated by a voltage close to the circuit's supply voltage. If the V_{in} voltage is lower than the V_{ref} voltage, the output will be LOW, or close to ground potential.

If the two input voltages are exactly equal, the output of the circuit will be equal to half the supply voltage.

There will be a very narrow range of in-between output values when the V_{in} voltage is very close to, but not quite equal to the V_{ref} voltage. This range of questionable voltages is very narrow, and can reasonably be ignored in most cases.

This leaves us with three possible input-output combinations:

Inputs	Output
$V_{in} > V_{ref}$	HIGH
$V_{in} = V_{ref}$	mid-point (zero)
$V_{in} < V_{ref}$	LOW

This circuit works because with no feedback resistor, the Norton amplifier's gain is very high. If there is just a small differential voltage (difference between the inverting and noninverting inputs), the Norton op amp will be saturated, and forced to either its positive or negative maximum output voltage. For more information on comparator circuits, refer to chapter 4.

UNDER-VOLTAGE INDICATOR

The circuit illustrated in Fig. 10-7 is a specialized comparator that gives a positive and unambiguous indication whenever a monitored input voltage drops below a specific level. A typical application for this circuit would be in portable, battery-operated equipment. When the batteries near the end of their useful lifespan, their voltage drops.

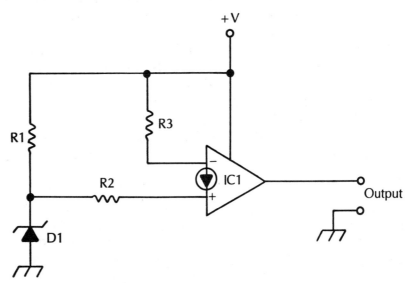

Fig. 10-7 Under-Voltage Indicator

This circuit can warn the user to replace the batteries before they go completely dead. When batteries go dead unexpectedly, it can often be disastrous. For example, if you want to record an important speech, you certainly don't want the batteries to go dead in the middle. A suitable parts list for this project appears as Table 10-6.

Table 10-6 Parts List for the Under-Voltage Indicator project of Fig. 10-7.

Part	Description
IC1	LM3900 quad Norton amplifier IC
D1	zener diode—see text
R1	10K 5 percent 1/4-watt resistor
R2	1 Megohm 5 percent 1/4-watt resistor
R3	2.2 Megohm 5 percent 1/4-watt resistor

The zener diode must be selected to suit the specific need. The under-voltage detector circuit will be triggered whenever the monitored input voltage, which also supplies power to the IC itself, drops below a value equal to twice the zener diode's breakover voltage. For example, if you use a 6.2-volt zener diode in this circuit, it will be triggered when the monitored voltage drops below about 12.5 (actually 12.4) volts.

When triggered, the circuit's output goes HIGH. As long as the monitored voltage is in the acceptable region, above the trigger point set by the zener diode, the circuit's output will be LOW.

The value of input resistor R2 also has an effect on the trip point of this circuit. Experiment with alternative values. This resistor should have a value of at least 750KΩ. You could use a potentiometer in series with a fixed resistor to manually adjust the trigger voltage of the circuit.

So far in this chapter, we have been working with circuits that control voltages. The Norton op amp's inputs are also designed to accept input currents. Therefore, the remainder of the projects in this chapter will be various current regulators, taking advantage of other characteristics of the Norton amplifier.

FIXED-CURRENT SOURCE

Ordinarily, the load determines how much current it will draw from the source in response to the voltages and resistances within the load, in accordance with Ohm's Law. In some applications, however, we want the current level to be independent of the load. In such applications, we need a *fixed-current source*, which always puts out a specific amount of current.

A fixed current source circuit built around an LM3900 Norton amplifier is illustrated in Fig. 10-8. The parts list for this project is given in Table 10-7.

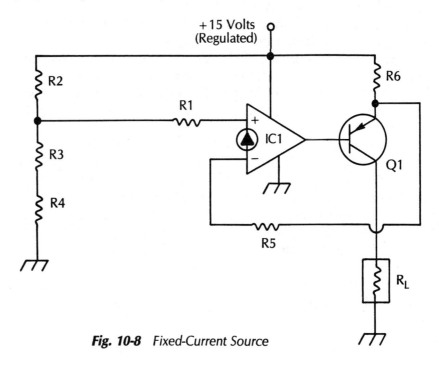

Fig. 10-8 Fixed-Current Source

Table 10-7 Parts List for the Fixed-Current Source project of Fig. 10-8.

Part	Description
IC1	LM3900 quad Norton amplifier IC
Q1	PNP transistor (2N3906, or similar)
R1, R5	1 Megohm 5 percent 1/4-watt resistor
R2, R6	1K 5 percent 1/4-watt resistor
R3	10K 5 percent 1/4-watt resistor
R4	3.9K 5 percent 1/4-watt resistor

This circuit could be considered a "current regulator," analogous to a voltage regulator. The output current is regulated, held to a specific value, and not permitted to fluctuate during normal operation.

Because the output current from this type of circuit is related to the voltage, the power supply should be well regulated. In the following discussion, we will assume that a +15 volt voltage regulator is being used to drive the circuit.

Using the component values from the parts list of Table 10-7, the output current from this circuit will be 1 mA.

The load, R1, being driven by the fixed current must be connected between the collector of output transistor Q1, and ground. The current source will hold constant as long as the load impedance is no greater than 14K. The load impedance may drop to zero ohms without affecting the current source.

The input to the Norton amplifier is derived from a simple voltage divider network made up of resistors R2, R3, and R4. The values of these resistors are selected to present a 14-volt input to the Norton op amp's noninverting input, through input resistor R1. For a high precision function, it would be a good idea to use 1 percent tolerance resistors in this voltage divider network. Resistors R3 and R4 could be replaced with a single 14K, 1 percent resistor.

In this circuit, feedback resistor R5 has the same value as input resistor R1, so we have a noninverting, unity gain amplifier:

$$G = R_5/R_1$$
$$= 1{,}000{,}000/1{,}000{,}000$$
$$= 1$$

The Norton amplifier automatically adjusts its output to provide an identical voltage at the junction of resistors R5 and R6. This voltage is placed on the emitter of transistor Q1, while the direct output of the Norton amplifier feeds the transistor's base. The collector is connected to the load.

Because there is +14 volts at the R5 end of resistor R6, and the full supply voltage (+15 volts) at the other end of this resistor, it necessarily follows that the voltage drop across this component is 1 volt. Knowing the value of resistor R6 permits us to use Ohm's Law to find the current flowing through it. According to the parts list, resistor R6 has a value of 1KΩ. In a critical case, use a high-precision 1 percent tolerance resistor here too.

The current flowing through resistor R6 in this circuit works out to:

$$I = E/R$$
$$= 1/1000$$
$$= 0.001 \text{ ampere}$$
$$= 1 \text{ mA}$$

This 1 mA current is derived from the emitter of transistor Q1. You should recall from your basic electronics theory that a transistor's emitter and collector currents are virtually identical, so the output current to the load, from the transistor's collector, is also about 1 mA, and is held at a constant value regardless of small to moderate fluctuations in the load impedance of R1.

The fixed output current from this circuit can be increased by decreasing the value of resistor R6. Reducing this resistance by half doubles the output current. Just use Ohm's Law to calculate the necessary value of resistor R6.

Do not try to make the output current too large, or the transistor or the IC could be damaged. Transistor Q1 may be almost any pnp type, such as the 2N3904. Just make sure that the transistor you select for this circuit can safely handle the desired output current. As a general rule, overrate the transistor's current handling capability by at least 20 to 30 percent.

SIMPLE CURRENT SINK

A *current sink* is just the opposite of a current source. A current sink draws a constant level of current. A simple current sink circuit built around the LM3900 Norton op amp is shown in Fig. 10-9, with the parts list appearing as Table 10-8.

In this circuit, a constant current will flow through load resistance R_1, which is connected between the collector of transistor Q1 and the circuit's positive supply voltage.

This current sink circuit is quite simple. Aside from the Norton amplifier, IC1, and the load, R_1, only two components are required; one resistor (R1), and an npn transistor (Q1). Almost any npn type, such as the 2N2222, or the 2N3906, may be used for transistor Q1 in this circuit. Just make sure that the transistor you select for this circuit can safely handle the desired output

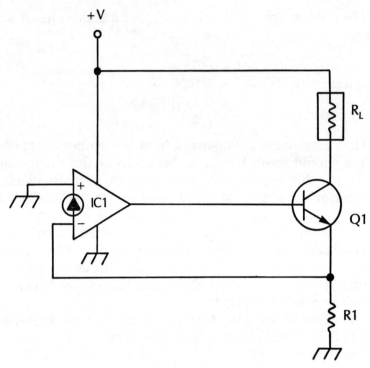

Fig. 10-9 Simple Current Sink

Table 10-8 Parts List for the Simple Current Sink project of Fig. 10-9.

Part	Description
IC1	LM3900 quad Norton amplifier IC
Q1	NPN transistor (2N2222, 2N3904, or similar)
R1	560 ohm 5 percent 1/4-watt resistor

current. Overrate the transistor's current handling capability by at least 20 to 30 percent.

The Norton amplifier's noninverting input is grounded in this circuit, seeing an effective input of zero. Meanwhile, the inverting input is receiving a feedback signal from the Norton amplifier's output through the base and emitter of transistor Q1. The voltage seen at the inverting input of the Norton op amp must obviously be equal to the voltage drop across resistor R1. This is equal to approximately 0.55 volt.

We can now use Ohm's Law to find the amount of current flowing through resistor R1:

$$I = E/R$$
$$= 0.55/560$$
$$= 0.0009821 \text{ ampere}$$
$$= 1 \text{ mA}$$

The current value is rounded off here. Component tolerances could easily account for as much error. In a very high-precision application, you could calculate the current more precisely and use a 1 percent tolerance resistor for R1.

This 1 mA of current through resistor R1 is taken from the emitter of transistor Q1. A transistor's emitter and collector currents are almost identical, so the output current to the load, from the transistor's collector, is also about 1 mA, and is held at a constant value, regardless of small to moderate fluctuations in the load impedance of R1.

The main difference between a current source circuit and a current sink circuit is the placement of the load, R1. In a current source circuit, the load is placed between the output and ground. In a current sink circuit, on the other hand, the load is placed between the output and V+, the positive supply voltage. In many applications, the two types of current regulator circuits are almost interchangeable, but certain uses may demand one type or the other.

IMPROVED CURRENT SINK

The current sink circuit of Fig. 10-9 is simple and reasonably effective, but it may not be good enough for some critical functions. An improved current sink circuit is illustrated in Fig. 10-10. A suitable parts list for this project appears as Table 10-9.

The current is better regulated in this sink circuit than in the preceding project, because a more precise input voltage is applied to the Norton amplifier. Zener diode D1, along with resistors R1 and R2, forms a simple voltage regulator network. Using the component values described in the parts list, Table 10-9, the input voltage applied to the Norton amplifier's noninverting input is fixed at +2.7 volts.

180 Voltage and Current Projects

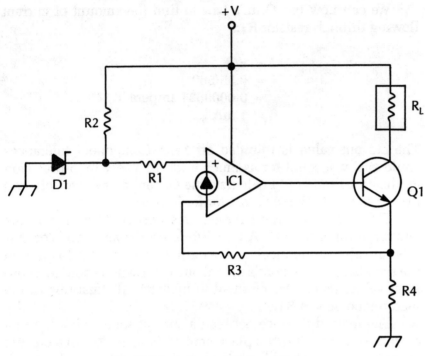

Fig. 10-10 Improved Current Sink

Table 10-9 Parts List for the Improved Current Sink project of Fig. 10-10.

Part	Description
IC1	LM3900 quad Norton amplifier IC
IC2	NPN transistor (2N2222, 2N3904, or similar)
D1	2.7 volt zener diode
R1, R3	1 Megohm 5 percent 1/4-watt resistor
R2	6.8K 5 percent 1/4-watt resistor
R4	2.7K 5 percent 1/4-watt resistor

Input resistor R1 and feedback resistor R3 have equal values (1 megohm), so the Norton op amp is set up as a unity gain non-inverting amplifier:

$$G = R_3/R_1$$
$$= 1{,}000{,}000/1{,}000{,}000$$
$$= 1$$

The amplifier's output is equal to its input. Consequently, 2.7 volts are dropped across resistor R4. Since this resistor has a value of 2.7K, Ohm's Law tells us that the current flowing through it is:

$$\begin{aligned} I &= E/R \\ &= 2.7/2700 \\ &= 0.001 \text{ ampere} \\ &= 1 \text{ mA} \end{aligned}$$

Other resistor values may be selected to give other current values. In a high-precision case, use a 1 percent tolerance resistor for R4.

This 1 mA current is derived from the emitter of transistor Q1. As the transistor's emitter and collector currents are nearly identical, the output current to the load, from the transistor's collector, is also about 1 mA, and is held at a constant value regardless of small-to-moderate fluctuations in the load impedance of R1.

Almost any npn type transistor, such as the 2N2222, or the 2N3906, may be used for Q1 in this circuit. Just make sure that the transistor you select for this circuit can safely handle the desired output current. Overrate the transistor's current handling capability by at least 20 to 30 percent.

For the best and most accurate results, it is strongly recommended that you operate this circuit with a well-regulated power supply.

❖ 11
Miscellaneous Norton Amplifier Projects

THIS FINAL CHAPTER WILL PRESENT TEN VARIED PROJECTS USING THE LM3900 Norton amplifier IC. Many of these projects don't have much in common with one another, but they are worth looking at, even though a specialized chapter is not justified. The varied projects offered here should give you a very good idea of the versatility of the Norton op amp.

FOUR-INPUT AND GATE

The Norton amplifier is a linear analog device, but in some applications, it can simulate digital circuitry. Figure 11-1 shows an analog AND gate circuit built around the LM3900. A suitable parts list for this simple project appears as Table 11-1.

Notice that there are four inputs and one output. The number of inputs may be varied to suit the use.

The output of the AND gate circuit goes HIGH when all of the inputs are HIGH. Under any other conditions, if one or more of the inputs is LOW, the output of the gate will also be LOW.

In discussing digital circuits, a *truth table* is used to concisely indicate all possible combinations of input states. Any input or output must be in one of two possible states: HIGH (1), or LOW (0). For a gating circuit with four inputs and a single output, as in our project here, there are 16 possible combinations:

Inputs A B C D	Output
0 0 0 0	0
0 0 0 1	0
0 0 1 0	0

Inputs A B C D	Output
0 0 1 1	0
0 1 0 0	0
0 1 0 1	0
0 1 1 0	0
0 1 1 1	0
1 0 0 0	0
1 0 0 1	0
1 0 1 0	0
1 0 1 1	0
1 1 0 0	0
1 1 0 1	0
1 1 1 0	0
1 1 1 1	1

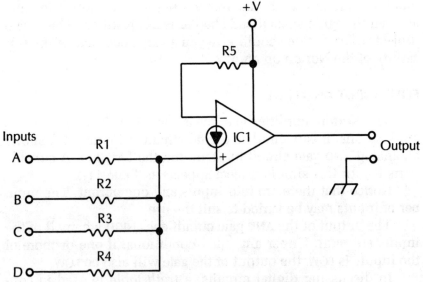

Fig. 11-1 Four-input AND Gate

Table 11-1 Parts List for the Four-input AND Gate project of Fig. 11-1

Part	Description
IC1	LM3900 quad Norton amplifier IC
R1, R2, R3, R4	2.2 Megohm 5 percent 1/4-watt resistor
R5	1 Megohm 5 percent 1/4-watt resistor

Notice that the output of a four input AND gate is HIGH if, and only if, input A, input B, input C, AND input D are all HIGH.

The values of input resistors R1 through R4 should be selected to suit the intended use. When all four inputs are HIGH, the current at the Norton amplifier's noninverting input should slightly exceed the fixed current through bias resistor R5 at the inverting input. If one or more of the four inputs are LOW, the noninverting current should be less than the inverting current.

When the AND gate output goes HIGH, or all four inputs are HIGH, the circuit's output voltage is close to the V+ supply voltage. When the AND gate output is LOW, or one or more of the inputs are LOW, the circuit's output voltage is close to ground potential.

This circuit is actually a variation on the basic voltage comparator. Refer to chapter 10.

A very simple but useful modification of this circuit would be to reverse the connections to the inverting and noninverting inputs. This will reverse the gate's input–output pattern:

Inputs A B C D	Output
0 0 0 0	1
0 0 0 1	1
0 0 1 0	1
0 0 1 1	1
0 1 0 0	1
0 1 0 1	1
0 1 1 0	1
0 1 1 1	1
1 0 0 0	1
1 0 0 1	1
1 0 1 0	1
1 0 1 1	1
1 1 0 0	1
1 1 0 1	1
1 1 1 0	1
1 1 1 1	0

This is a NAND, or "Not AND" gate. The output is HIGH *unless* all four inputs are HIGH. If input A, input B, input C, and input D are all HIGH, then the circuit output is LOW.

FLIP-FLOP

Our next project is another pseudo-digital circuit. Figure 11-2 shows a *flip-flop*, or *bistable multivibrator* circuit. The term

186 Miscellaneous Norton Amplifier Projects

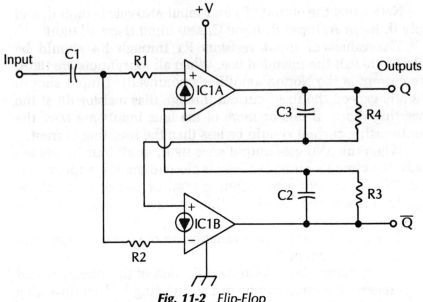

Fig. 11-2 Flip-Flop

"bistable" means that there are two stable output states from this circuit; HIGH, near V+; and LOW, near ground potential. No other output conditions are possible with this type of circuit.

Either output state may be held indefinitely as long as power is applied to the circuit. The output reverses its state from LOW to HIGH or from HIGH to LOW each time a trigger pulse is detected at the circuit's input. The output "flip-flops" back and forth on successive input pulses. The operation of a flip-flop circuit is illustrated in Fig. 11-3.

A suitable parts list for this project appears as Table 11-2. None of the component values are particularly critical in this circuit. Notice that two stages of a LM3900 quad Norton amplifier IC are used in this circuit.

Notice also that the circuit actually has two outputs, labeled "Q" and "$\overline{Q}$." The bar over the second Q is read as "Not." It indicates an inverted output. The two outputs of this circuit are always at opposite states. When Q is LOW, $\overline{Q}$ (or Not Q) is HIGH. Similarly, when Q goes HIGH, $\overline{Q}$ (or Not Q) goes LOW.

LOW-TEMPERATURE ALARM

The circuit shown in Fig. 11-4 is a modified voltage comparator, being used here as a thermal monitor. When the tempera-

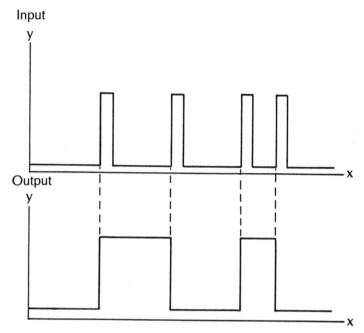

Fig. 11-3 *The output of a Flip-flop reverses states each time an input trigger pulse is received.*

Table 11-2 Parts List for the Flip-Flop project of Fig. 11-2.

Part	Description
IC1	LM3900 quad Norton amplifier IC
C1, C2, C3	0.001 µF capacitor
R1, R2	100K 5 percent ¼-watt resistor
R3, R4	1 Megohm 5 percent ¼-watt resistor

ture sensed by thermistor R1 drops below a specific, pre-set value, the circuit's output goes HIGH. As long as the monitored temperature is higher than the critical value, the circuit output is LOW. A suitable parts list for this project is given in Table 11-3.

Basically, this circuit is another variation on the basic voltage comparator (refer to chapters 4 and 10). The input voltage to the noninverting input of the Norton amplifier is set by a simple voltage divider network made up of thermistor R1 and potentiometer R2.

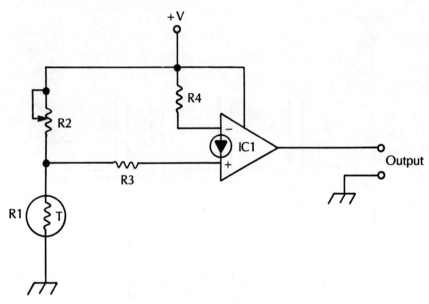

Fig. 11-4 Low-temperature Alarm

**Table 11-3 Parts List for the
Low-temperature Alarm project of Fig. 11-4.**

Part	Description
IC1	LM3900 quad Norton amplifier IC
R1	thermistor
R2	10K potentiometer
R3	1 Megohm 5 percent 1/4-watt resistor
R4	2.2 Megohm 5 percent 1/4-watt resistor

The thermistor in this circuit should be a negative temperature coefficient (NTC) type. The resistance of the component decreases with increases in its temperature.

Assuming that the setting of potentiometer R2 is not altered, the voltage fed into the noninverting input through resistor R3 will vary as the resistance of thermistor R1 changes in response to temperature. If the temperature increases, the resistance of R1 will drop, decreasing the input voltage.

As the monitored temperature decreases, the resistance of thermistor R1 goes up, increasing the input voltage, and thus the current seen at the Norton amplifier's noninverting input.

If the current at the inverting input is greater than the current at the noninverting input, the circuit's output will go LOW. If the noninverting current exceeds the inverting current, the circuit output will go HIGH, indicating that the monitored temperature has dropped below the preset cutoff value. The output can be used to drive some sort of alarm device, or it can turn on a furnace or heating element through a relay of some sort.

By adjusting the resistance of potentiometer R2 we can change the necessary thermistor (R1) resistance needed to trip the circuit. In effect, potentiometer R2 controls the sensitivity of the circuit, setting the minimum temperature the thermistor can sense without the output going HIGH.

HIGH-TEMPERATURE ALARM

This project is just the opposite of the preceding one. In the last circuit, the output went HIGH when the monitored temperature was too low. In this circuit, shown in Fig. 11-5, the output goes HIGH when the monitored temperature exceeds a specific, preset value.

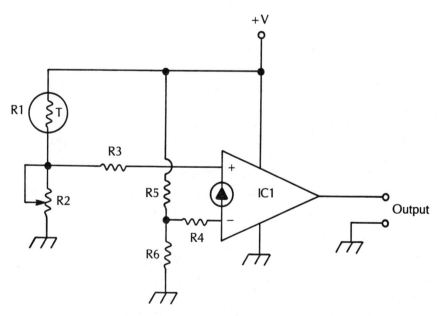

Fig. 11-5 High-temperature Alarm

The main differences between this circuit and the one illustrated in Fig. 11-4 are the relative positions of thermistor R1, and the sensitivity potentiometer, R2. The voltage divider made up of these two components supplies an input voltage through resistor R3 to the Norton amplifier's noninverting input.

A second voltage divider network made up of fixed resistors R5 and R6 feeds a constant reference voltage into the inverting input through resistor R4. A suitable parts list for this project appears as Table 11-4.

Table 11-4 Parts List for the
High-temperature Alarm project of Fig. 11-5.

Part	Description
IC1	LM3900 quad Norton amplifier IC
R1	thermistor
R2	10K potentiometer
R3, R4	1 Megohm 5 percent 1/4-watt resistor
R5, R6	10K 5 percent 1/4-watt resistor

The thermistor in this circuit should be an NTC, type. The resistance of the component decreases with increases in its temperature.

Assuming that the setting of potentiometer R2 is not altered, the voltage fed into the noninverting input through resistor R3 will vary as the resistance of thermistor R1 changes in response to temperature. If the temperature increases, the resistance of R1 will drop, increasing the input voltage. Likewise, as the monitored temperature decreases, the resistance of R1 goes up, decreasing the input voltage, and thus the current seen at the Norton amplifier's noninverting input.

If the current at the inverting input is greater than the current at the noninverting input, the circuit's output will go LOW. If the noninverting current exceeds the inverting current, the circuit output will go HIGH, indicating that the monitored temperature has gone above the preset cutoff value. The output can be used to drive some sort of alarm device, or it can turn on a fan or air conditioning device through a relay of some sort.

By adjusting the resistance of potentiometer R2 we can change the necessary thermistor (R1) resistance needed to trip the circuit. In effect, potentiometer R2 controls the sensitivity of the circuit, setting the maximum temperature the thermistor can sense without the output going HIGH.

SQUARE-WAVE GENERATOR

Earlier in this chapter, we worked with a flip-flop, or bistable multivibrator circuit. A related type of circuit is the *astable multivibrator*, which is essentially a flip-flop circuit with positive feedback, forcing it into oscillation.

Like the bistable multivibrator, the astable multivibrator has two possible output states—HIGH and LOW. There are no intermediate output values. In a practical circuit, the transition time between output states is short enough to be considered negligibile.

In a bistable multivibrator, both output states are stable (can be held indefinitely), but neither output state is stable in the astable multivibrator. If the output is HIGH, this state will be held for a specific period of time, determined by component values within the circuit and then will automatically switch into the LOW state.

The LOW state will also be held for a specific period of time, determined by component values before automatically switching back into the HIGH output state. The output continues to switch back and forth between these two states at a regular rate as long as power is applied to the circuit. No external input signal is required.

A practical astable multivibrator circuit built around the LM3900 Norton op amp is illustrated in Fig. 11-6. A suitable parts list for this project is given in Table 11-5.

A typical output signal from this circuit is shown in Fig. 11-7. Notice that this is a repeating (ac) waveform. It is HIGH for exactly half of each complete cycle, and LOW for the other half. This waveshape is rather squarish, so it is naturally known as a *square wave*. A square wave is a special form of the *rectangular wave* with a duty cycle of 1:2.

Another name for an astable multivibrator is a *square-wave generator*. This more descriptive name is probably more useful.

This circuit is fairly simple. In addition to the Norton op amp IC itself, only five passive components are required—four resistors and a capacitor. All five of these component values have

192 Miscellaneous Norton Amplifier Projects

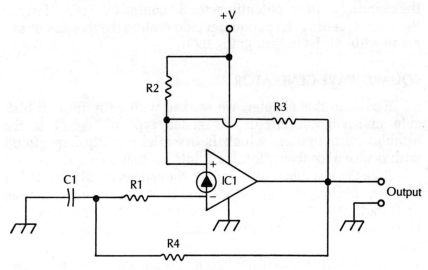

Fig. 11-6 Square-Wave Generator

Table 11-5 Parts List for the Square-Wave Generator project of Fig. 11-6.

Part	Description
IC1	LM3900 quad Norton amplifier IC
C1	0.02 µF capacitor
R1	3.3 Megohm 5 percent 1/4-watt resistor
R2, R3	10 Megohm 5 percent 1/4-watt resistor
R4	33K 5 percent 1/4-watt resistor

an effect on the output frequency, especially the capacitor and resistor R4.

Using the component values suggested in the parts list (Table 11-5), the output frequency is approximately 1 kHz. You are encouraged to breadboard this circuit and experiment with different component values. This circuit is functional over a very wide range of output frequencies up to several kiloHertz.

The LM3900 Norton amplifier is not very well suited to high frequency applications because of its limited slew rate—about 0.5 volt/microsecond. The slower the slew rate of the amplifier used in a multivibrator circuit, the longer the transition time when switching between output states. The LM3900's slew rate is good enough for most audio frequency and subaudio frequency uses.

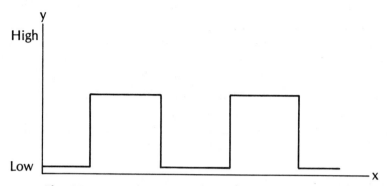

Fig. 11-7 *A square wave is* HIGH *for half of each cycle.*

Let's take a look at what is happening within this circuit. Let's begin by assuming that the output is currently in its HIGH state, near V+. Resistors R2 and R3 are effectively in parallel. Notice that these two resistors have equal values.

As soon as the Norton amplifier's output goes HIGH, capacitor C1 starts to charge from the output voltage through feedback resistor R4. The voltage across capacitor C1 is applied to the Norton amplifier's inverting input through resistor R1.

At some point the voltage across capacitor C1 will exceed two-thirds of the circuit's supply voltage. This forces the Norton amplifier to switch its output into the LOW state, virtually ground potential.

As soon as the output goes LOW, resistor R3 is effectively switched out of the circuit. Only resistor R3 controls the current to the Norton op amp's noninverting input.

Now that the output voltage is practically zero, LOW, the current through feedback resistor R4 changes direction. Capacitor C1 starts to discharge through resistor R4.

At some point, the current through inverting input resistor R1 will be less than the current through noninverting input resistor R3. Using the component values from parts list of Table 11-5, this will occur when the voltage across capacitor C1 drops below one-third of the circuit's supply voltage. When this happens, the Norton amplifier switches its output state again. The output returns to the HIGH state, and the entire cycle is repeated for as long as power is applied to the circuit.

The RC time constant of capacitor C1 and resistor R4 is the primary factor in determining the circuit's output frequency, but resistors R2 and R3 also have a significant effect because they set the circuit's switchover voltages.

In experimenting with this circuit, you will probably get the best results by keeping the parts list values for resistors R1, R2, and R3, and substituting other values for resistor R4 and capacitor C1. However, it may be informative for you to check what happens when the values of resistors R1, R2, or R3 are altered. You may well want to see how the circuit's operation is affected when resistors R2 and R3 do not have equal values.

ALTERNATIVE SQUARE-WAVE GENERATOR

A different and somewhat more sophisticated square-wave generator circuit is shown in Fig. 11-8. A typical parts list for this project appears as Table 11-6.

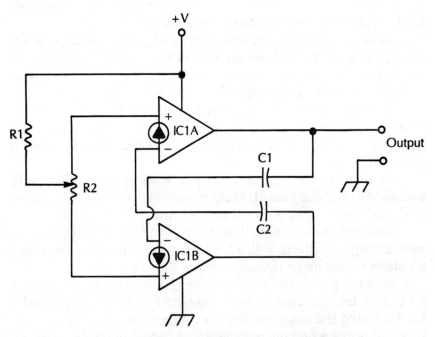

Fig. 11-8 Alternate Square-Wave Generator

Notice that two sections of an LM3900 quad Norton amplifier IC are used in this circuit. This circuit more closely resembles the flip-flop, or bistable multivibrator circuit presented earlier in this chapter.

Table 11-6 Parts List for the Alternate Square-Wave Generator project of Fig. 11-8.

Part	Description
IC1	LM3900 quad Norton amplifier IC
C1, C2	0.1 µF capacitor
R1	10K 5 percent ¼-watt resistor
R2	500K potentiometer

While only one output, from IC1A, is shown in this circuit diagram, a second, complementary NOT output can be taken from the output of IC1B. This second output will also have the opposite state as the primary output. That is, when the output of IC1A is HIGH, the output of IC1B will be LOW. Similarly, when the output of IC1A is LOW, the output of IC1B will be HIGH. This secondary output may be useful in certain applications.

When the output of IC1A is HIGH, capacitor C1 is charged until its current into the inverting input of IC1B forces the Norton amplifiers to reverse states. While the output of IC1A is LOW, the output of IC1B is HIGH, charging capacitor C2, which feeds a current into the inverting input of IC1A. At some point, this current will be sufficient to force the two Norton amplifiers to reverse output states again, and the entire cycle is repeated, as long as power is applied to the circuit.

The capacitor charging rate is set by potentiometer R2. By altering the capacitor charging rate with this control, the length of each cycle is changed; therefore, potentiometer R2 directly controls the circuit's output frequency.

If your individual use does not call for a manually variable output frequency, you might want to replace potentiometer R2 with a screwdriver adjust trimpot, or possibly with a pair of fixed resistors of suitable values.

Once again, you are encouraged to experiment with component values in this circuit. If capacitors C1 and C2 are given unequal values, the HIGH and LOW times will be different, giving a rectangular wave with a duty cycle not equal to 1:2. Of course, the output signal will no longer be a true square wave with a different duty cycle. Changing the duty cycle of the waveform in this simple circuit will also alter the output frequency.

NARROW-WIDTH PULSE GENERATOR

A square wave, as we've said earlier, is a specialized form of a rectangular wave, with a duty cycle of 1:2. A rectangular wave can have any duty cycle. The duty cycle of the rectangular wave is a comparison ratio of the waveshape's HIGH time to the time of the total cycle. The duty cycle defines the signal's harmonic content.

Another specialized form of rectangular wave is the *pulse wave*. This is a rectangular wave with a duty cycle that results in a very short HIGH time per cycle. Figure 11-9 shows a circuit for generating pulse waves. A suitable parts list for this project is given in Table 11-7.

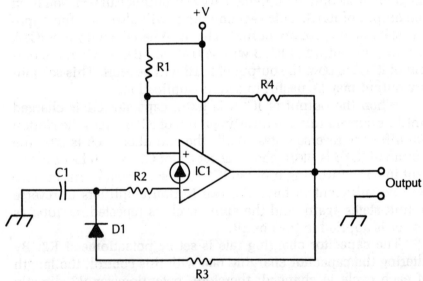

Fig. 11-9 *Narrow-width Pulse Generator*

Table 11-7 Parts List for the Narrow-width Pulse-Generator project of Fig. 11-9.

Part	Description
IC1	LM3900 quad Norton amplifier IC
D1	diode (1N914, or similar)
C1	0.02 uF capacitor
R1, R4	10 Megohm 5 percent 1/4-watt resistor
R2	2.2 Megohm 5 percent 1/4-watt resistor
R3	33K 5 percent 1/4-watt resistor

A pulse wave, like the signal generated by this project, is illustrated in Fig. 11-10. This sketch is somewhat exaggerated. Actually, the HIGH portion of the cycle is much shorter than indicated here. Using the component values suggested in the parts list (Table 11-7), the duty cycle is approximately 1:60. Of course, you are encouraged to experiment with other component values in this circuit.

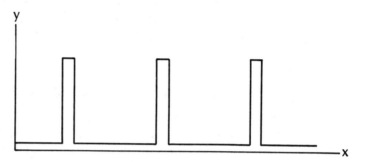

Fig. 11-10 *A typical output signal from the circuit of Fig. 11-9*

Table 11-8 Parts List for the Variable Duty-Cycle Rectangular-Wave Generator project of Fig. 11-12.

Part	Description
IC1	LM3900 quad Norton amplifier IC
D1, D2	diode (1N914, or similar)
C1	0.01 µF capacitor
R1, R7	10 Megohm 5 percent ¼-watt resistor
R2	3.3 Megohm 5 percent ¼-watt resistor
R3	8.2K 5 percent ¼-watt resistor
R4, R5	1K 5 percent ¼-watt resistor
R6	100K potentiometer

Because of the wide HIGH-to-complete-cycle ratio, the harmonic content of this waveform is very strong, with few of the natural harmonics missing from the signal.

Compare the narrow width pulse generator circuit of Fig. 11-8 with the first square-wave generator circuit shown in Fig. 11-6. You should be able to see that there is a great deal of similarity between the two circuits. In this project, the most important difference lies in the inclusion of diode D1.

Capacitor C1 is charged during the HIGH portion of the output waveform, and discharged during the LOW portion of the cycle. Because diode D1 permits current to flow in one direction, but not the other, the capacitor is charged through feedback resistor R3, and discharged through input resistor R2.

The value of input resistor R2 is much larger than the value of feedback resistor R3. This means that the LOW portion of the cycle has a significantly larger RC time constant than the HIGH portion of the cycle. The output frequency of this circuit is determined primarily by the values of capacitor C1, and by resistors R2 and R3.

VARIABLE DUTY-CYCLE RECTANGULAR-WAVE GENERATOR

Rectangular waves can be generated with many differing duty cycles, as illustrated in Fig. 11-11. Remember that the duty cycle is a measurement of how much of each complete cycle is in the HIGH state. A square wave is a special case of a rectangular wave. It is a symmetrical rectangle. That is, the LOW time exactly equals its HIGH time.

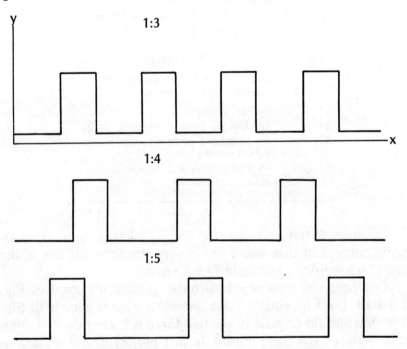

Fig. 11-11 Rectangular waves may have a variety of different duty cycles.

The duty cycle figure also defines the harmonic content of the waveform. Any repeating ac waveform is made up of a fundamental frequency, which is the nominal frequency of the signal as a whole, and additional frequency components known as *harmonics*. The one exception is the sine wave, which consists of the fundamental only. Each harmonic is a whole number multiple of the fundamental frequency. For example, if the fundamental frequency is 100 Hz, the lower harmonics, usually the strongest, will be as follows:

100 Hz	Fundamental	
200 Hz	Second Harmonic	× 2
300 Hz	Third Harmonic	× 3
400 Hz	Fourth Harmonic	× 4
500 Hz	Fifth Harmonic	× 5
600 Hz	Sixth Harmonic	× 6
700 Hz	Seventh Harmonic	× 7
800 Hz	Eighth Harmonic	× 8
900 Hz	Ninth Harmonic	× 9
1000 Hz	Tenth Harmonic	× 10
1100 Hz	Eleventh Harmonic	× 11
1200 Hz	Twelfth Harmonic	× 12

and so forth.

Most practical waveforms, including all rectangular waves, do not include all possible harmonics. Some are omitted. The difference between various waveforms lies in which harmonics are included and which are omitted, and the relative amplitudes of the various frequency components.

The duty cycle of a rectangular wave has a direct connection to the signal's harmonic content. A square wave, with a duty cycle of 1:2, has all harmonics except those which are evenly divisible by two. A square wave consists of the following frequency components:

Fundamental
Third Harmonic
Fifth Harmonic
Seventh Harmonic
Ninth Harmonic
Eleventh Harmonic

and so forth.

A rectangular wave with a duty cycle of 1:3 contains all of the harmonics, except for those which are exact whole number multiples of three:

> Fundamental
> Second Harmonic
> Fourth Harmonic
> Fifth Harmonic
> Seventh Harmonic
> Eighth Harmonic
> Tenth Harmonic
> Eleventh Harmonic

and so forth.

If a rectangular wave has a duty cycle of 1:4, every fourth harmonic will be missing from the output waveform:

> Fundamental
> Second Harmonic
> Third Harmonic
> Fifth Harmonic
> Sixth Harmonic
> Seventh Harmonic
> Ninth Harmonic
> Tenth Harmonic
> Eleventh Harmonic

and so forth.

If the duty cycle of the rectangular wave is 1:5, all harmonics that are whole number multiples of five will be omitted:

> Fundamental
> Second Harmonic
> Third Harmonic
> Fourth Harmonic
> Sixth Harmonic
> Seventh Harmonic
> Ninth Harmonic
> Eleventh Harmonic
> Twelfth Harmonic

and so forth. Different applications may well call for rectangular waves of varying duty cycles.

The rectangular wave generator circuit of Fig. 11-12 permits manual control over the duty cycle of the output waveform. A suitable parts list for this project appears as Table 11-8. The unipolar conduction of diodes is used to unequally charge and discharge capacitor C1 in this circuit.

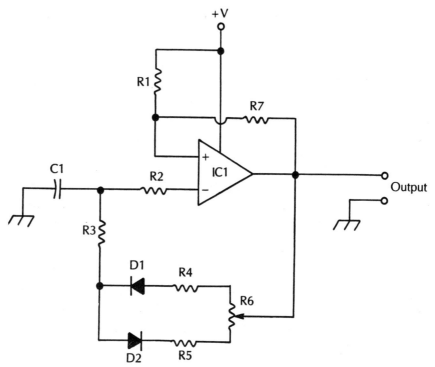

Fig. 11-12 Variable Duty-Cycle Rectangular-Wave Generator

During the HIGH portion of the cycle, diode D1 is forward biased, and the capacitor is charged through resistor R4 and the upper half of potentiometer R6. Diode D2 is reverse biased, preventing current from flowing through resistor R5 and the lower half of potentiometer R6.

When the output is in the LOW part of the cycle, the situation is reversed. Now diode D1 is reverse biased, blocking current flow through resistor R4 and the upper half of potentiometer R6. At the same time, diode D2 is forward biased, so the capacitor is permitted to discharge through resistor R5 and the lower half of potentiometer R6.

The RC time constant for the HIGH portion of the cycle is controlled by the values of capacitor C1, resistor R4 and the upper half of potentiometer R6, while the RC time constant for the LOW portion of the cycle is controlled by the values of capacitor C1, resistor R5 and the lower half of potentiometer R6.

If resistors R4 and R5 are given equal values, the duty cycle of the output waveform will be 1:2 (a square wave) when potentiometer R6 is set to the exact midpoint of its range. Changing the duty cycle alters the overall timing of the cycle, so the output frequency will also be changed.

This is a good circuit to breadboard and experiment with. Try different component values for all of the passive components (the resistors, and capacitor C1) and check out the results.

FUNCTION GENERATOR

A *function generator* is a signal generator circuit that can put out more than a single waveform. A function generator circuit built around two sections of an LM3900 quad Norton amplifier IC is illustrated in Fig. 11-13. A suitable parts list for this project is given in Table 11-9. Experiment with alternative component values.

This circuit has two outputs, which can be used simultaneously or separately. Both output signals are always at the exact same frequency. The difference between the outputs is in the waveshape. One output is a rectangular wave. Using the component values suggested in the parts list, the duty cycle is about

Table 11-9 Parts List for the Function Generator project of Fig. 11-13.

Part	Description
IC1	LM3900 quad Norton amplifier IC
D1	diode (1N914, or similar)
C1	0.01 µF capacitor
C2	0.047 µF capacitor
R1	100K 5 percent ¼-watt resistor
R2	33K 5 percent ¼-watt resistor
R3	1K 5 percent ¼-watt resistor
R4, R5	1 Megohm 5 percent ¼-watt resistor
R6	10K 5 percent ¼-watt resistor

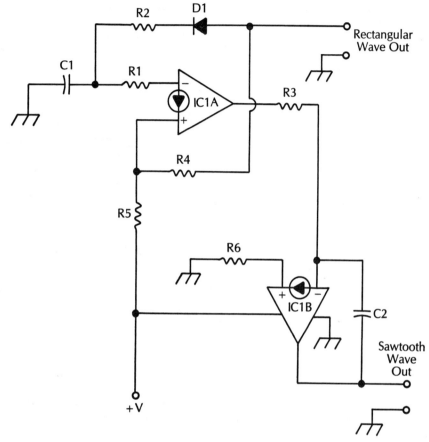

Fig. 11-13 *Function Generator.*

1:4. The second output generates a sawtooth wave. These two output waveforms from this circuit are illustrated in Fig. 11-14.

Essentially, IC1A, along with its associated components, is a simple rectangular-wave generator circuit, similar to the earlier circuits of Figs. 11-6 and 11-12.

IC1B, and its associated components, then, convert the rectangular wave into a sawtooth wave, which includes all of the harmonics, both even and odd.

DUAL LED FLASHER

LED flasher projects are not overly useful, but they are always popular. It is fascinating to watch them in action. They

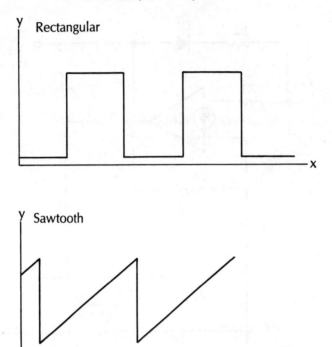

Fig. 11-14 *These are the output signals generated by the circuit of Fig. 11-13.*

can be put to work in some practical circuits, such as eye-catching displays or warnings, but mostly, they are just for fun projects.

The circuit shown in Fig. 11-15 is a dual LED flasher. It blinks two LEDs on and off. A typical parts list for this project is given in Table 11-10.

The two LEDs, D1 and D2, are always in opposite states. As long as power is applied to the circuit, one LED is lit, and the other is dark. These conditions keep switching back and forth. That is, if we start out with LED D1 on and LED D2 off, the circuit will soon switch LED D1 off and LED D2 on. Then, after a short time, LED D1 will be turned back on and LED D2 will be turned back off.

Basically, this circuit is a low-frequency square-wave generator. The actual frequency or flash rate can be manually controlled by potentiometer R3.

By all means, experiment with other component values in this circuit. The basic circuit is very similar to the alternative

Dual LED Flasher   205

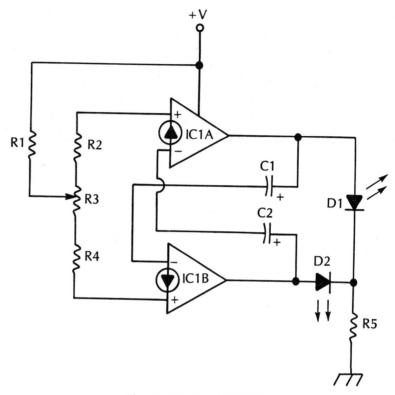

Fig. 11-15 Dual LED Flasher

Table 11-10 Parts List for the Dual LED Flasher project of Fig. 11-15.

Part	Description
IC1	LM3900 quad Norton amplifier IC
D1, D2	LED
C1, C2	22 μF 35 Volt electrolytic capacitor
R1	10K 5 percent 1/4-watt resistor
R2, R4	1K 5 percent 1/4-watt resistor
R3	500K potentiometer
R4	680 ohm 5 percent 1/4-watt resistor

square-wave generator project shown back in Fig. 11-8. If capacitors C1 and C2 are given unequal values, the two LEDs will not be lit for equal time periods.

If the circuit's switching frequency is made too high, both LEDs will appear to be continuously lit, although possibly at less than normal intensity. Actually they are still alternately blinking on and off, but at a rate too fast for the human eye to detect the individual flashes. This phenomenon is known as the "persistance of vision," and is the reason movies, which are actually a series of closely spaced still photographs, appear to move.

After going through the two dozen Norton amplifier projects in this section and the two dozen transconductance amplifier projects in the first section of this book, you should be able to come up with many additional applications for these devices on your own.

Happy experimenting.

Index

A

ac coupling, 37
ac-coupled inverting amplifier, 36-40
   I-bias and gain in, 37, 38
   open-loop configuration, 39
   voltage divider, 38
AGC amplifier, 44-46
   I-bias in, 46
   LM13600 OTA in, 45
   negative feedback system in, 46
alarm
   high-temperature, Norton amplifier, 189-191
   low-temperature, Norton amplifier, 186-189
amplifier, 31-53
   AGC, 44-46
   attenuation in, 31
   buffers, 31
   current-differencing, 125, 137
   definition of, 31
   differential, CA3080 OTA in, 16, 18, 19
   differential, direct-coupled, 32
   differential, Norton amplifiers, 151-155
   differential, op amps in, 7, 9
   differential, transconductance op amps in, 9, 31, 32
   gain in, 31
   inverting, ac-coupled, 36-40
   inverting, low-power, 40-42
   inverting, Norton amplifiers, 145-148
   inverting, op amp in, 5
   inverting, transconductance op amps in, 31
   inverting, variable gain, 43-44
   noninverting, Norton amplifiers, 148-151
   noninverting, op amp in, 6
   noninverting, transconductance op amps in, 31, 32
   Norton amplifier, 145-161
   voltage controlled, 11, 47-53
   voltage differencing, 137
   wide-bandwidth/high-gain, 155-157
amplitude modulation, 75
   four-quadrant multiplier, 82, 84-87
   phase-amplitude modulator, 87-89
   ring modulator, 82-84
   two-quadrant multiplier, 75
amplitude modulator
   CA3080, 75-77
   CA3080, I-bias and gain, 77
   CA3080, resistor values, 77
   LM13600, 78-79
   sideband generation, 79-82
   voltage-controlled amplifier vs., 76
analog switch, 55-58
   ON/OFF states, 55-57
   resistor values, 57
   sample-and-hold circuits as, 67-70
   SPST switch action vs., 57
   VCA vs., 55
analog-to-digital conversion, sample-and-hold circuit for, 70
AND gate, four-input, 183-185

Index

astable multivibrator, 91-94, 191-194
attenuation, 6
   amplifier, 31
automation
   voltage-controlled amplifier for, 47
   voltage-controlled resistance and, 108-110

B

band-pass filters, 112, 114, 158
band-reject filters, 113, 114, 158
bandwidth
   CA3080 OTA, 14, 39
   LM3900 Norton amplifier, 136
   wide-bandwidth/high-gain amplifier, 155-157
battery supplies, under-voltage indicator for, 173-174
biasing (see also I-bias), Norton amplifier, 127-128
bistable multivibrator (see flip-flop)
buffers, 31
   Darlington transistors for, 24, 25, 26
   impedance matching with, 31
   LM13600 dual OTA, 24, 25, 26
   load isolation with, 31
   op amps in, 7
   polarity inversion with, 31
   sample-and-hold circuit, 68-69
   unity gain in, 31

C

CA3080 OTA, 13-24
   ac-coupled inverting amplifier, 36-40
   amplitude modulator, 75-77
   analog switch, 55-58
   bandwidth of, 14, 39
   block diagram of, 22
   common mode rejection ratio, 14
   compensation, 21-24
   current mirrors (CM) in, 16-21
   differential amplifier in, 16, 18, 19
   direct-coupled differential amplifier using, 32
   fast inverting switch using, 58-60
   forward transconductance rating, 14
   four-quadrant multiplier, 84-87
   harmonics remover circuit, 105-108
   I-bias, 14
   impedance in, 15
   internal structure of, 16-21
   inverting voltage comparator, 71-73
   low-power inverting amplifier, 40-42
   negative feedback circuit using, 40
   noise in, excessive slew/bandwidth, 39
   noninverting voltage comparator, 73-75
   phase-amplitude modulator, 87-89
   precision current source, 110-111
   random music maker, 100-103
   sample-and-hold circuit using, 67-70
   Schmitt trigger, 60-67
   short-circuiting feature of, 14
   slew rate, 16, 39, 59
   specifications for, 13-16
   square-wave generator, 91-94
   variable duty-cycle rectangle wave generator, 94-97
   voltage-controlled amplifier using, 47
   voltage output range, 15-16
capacitors, sample-and-hold circuits, 69
carrier signal, 55, 75, 80-81
clipping, op amps, 23
clock, sample-and-hold circuit, 69
common mode rejection ratio, CA3080 OTA, 14
comparator
   fast inverting switch as, 59
   reference voltages in, 71
   voltage, inverting, 71-73
   voltage, noninverting, 73-75
   voltage, Norton amplifier, 133, 171-173
   Schmitt trigger as, 61
compensation
   CA3080 OTA, 21-24
   LM3900 Norton amplifier, 136
control pulse generator, sample-and-hold circuit, 69-70
conversions
   analog to pulse waveforms, Schmitt trigger for, 61, 62
   triangle-to-sine wave, 11
coupling
   ac, 37
   direct, 32

Index

current differencing amplifier, 137
current drain, 40
current mirrors (CM)
  CA3080 OTA, 16-21
  current sink usage of, 19
  current source usage of, 20
  LM3900 Norton amplifier, 142-143
current sinks
  current mirrors (CM) used as, 19
  Norton amplifier, 177-181
  Norton amplifier, improved version, 179-181
  precision, Norton amplifier for, 133
current source
  current mirrors (CM) used as, 20
  fixed, 174-177
  precision, 110-111
  precision, Norton amplifier for, 133
current-differencing amplifier, 125
cutoff frequency, filters, 112, 115, 159, 160, 161

D

Darlington transistors
  CA3080 OTA, 20
  LM13600 dual OTA, 24, 25, 26
data recording, sample-and-hold circuit for, 70
dead zones, 65, 170
decibels, gain, conversion to, 35
  transconductance op amps in, 32
differential amplifier
  CA3080 OTA using, 16, 18, 19
  direct-coupled, 32
  Norton amplifier in, 151-155
  op amp in, 7, 9
  polarity in, 32
  transconductance op amps in, 9, 31, 32
direct coupling, 32
direct-coupled differential amplifier, 32
  gain and I-bias in, 32, 35
  open-loop configuration, 36
  voltage divider, 34
distortion, positive feedback and oscillation, 91
dual LED flasher, 203-206
duty cycle, 94, 95, 196
  fundamentals, 199, 200
  harmonics, 199, 200
  variable duty-cycle rectangle wave generator, 94-97, 198-202

F

fast inverting switch, 58-60
  comparator function of, 59
  I-bias and gain in, 59
  resistor values, 59
  self-automated switching function in, 59
feedback
  negative (see negative feedback)
  positive, 91
filters, 111-119
  band-pass, 112, 114, 158
  band-reject, 113, 114, 158
  cutoff frequency, 112, 115, 159, 160, 161
  fundamentals and harmonics, 115-117
  harmonics remover circuit vs., 105-108
  high-pass, 112, 113
  high-pass, Norton amplifier for, 157-161
  high-pass, voltage controlled, 119-121
  low-pass, 112, 114, 158
  notch, 113
  voltage controlled, 11, 105
  voltage controlled, low-pass, 111-119
fixed-current source, 174-177
flasher, LED, dual, 203-206
flip-flops, Norton amplifier, 185-186, 191
four-input AND gate, 183-185, 183
four-quadrant multipliers, 82, 84-87
  calibration of, 87
frequency, 80, 81
  cutoff (see cutoff frequency; filters)
  fundamentals, 80, 81, 115, 199, 200
  harmonics, 80, 81, 115, 199, 200
  harmonics remover circuit, 105-108
frequency compensation
  LM3900 Norton amplifier, 136
  op amps, 23
  RC network for, 23
  transconductance op amps, 24

frequency modulation, 75
function generator, 202-203
fundamentals, 80, 81, 95, 115, 199, 200

G

gain, 3
  ac-coupled inverting amplifier, 37, 38
  amplifier, 31
  amplitude modulator, CA3080, 77
  automatic control (AGC) amplifier, 44-46
  CA3080 OTA, 14
  decibel conversion of, 35
  direct-coupled differential amplifier, 32, 35
  fast inverting switch, 59
  inverting voltage comparator, 72-73
  LM13600 dual OTA, 29
  LM3900 Norton amplifier, 139-140
  low-power inverting amplifier, 40, 41
  negative (attenuation), 6
  Norton amplifier, 130, 132
  op amps, 4, 6, 7, 8
  ring modulator, 83
  transconductance op amps, 9, 10
  unity (*see* unity gain)
  variable duty-cycle rectangle wave generator, 94-97, 198-202
  variable-gain inverting amplifier, 43-44
  wide-bandwidth/high-gain amplifier, 155-157
gates, 183
  truth tables, 183

H

harmonics, 80, 81, 95, 115, 199, 200
harmonics remover, 105-108
high-pass filter, 112, 113
  Norton amplifier for, 157-161
high-pass voltage-controlled filter, 119-121
high-temperature alarm, Norton amplifier, 189-191
hysteresis
  dead zones and, 65, 170
  noise vs., 65, 170
  Schmitt trigger and, 65, 170
  Schmitt trigger, variable, 64-67

I

I-bias, 4, 9, 10, 12
  ac-coupled inverting amplifier, 37, 38
  AGC amplifier, 46
  amplitude modulator, CA3080, 77
  CA3080 OTA, 14
  direct-coupled differential amplifier, 32, 35
  fast inverting switch, 59
  LM13600 dual OTA, 28, 29
  low-power inverting amplifier, 42
  noninverting voltage comparator, 74
  sample-and-hold circuits, 69
  Schmitt trigger, 62
  variable duty-cycle rectangle wave generator, 97
  variable-hysteresis Schmitt trigger, 65-66
  voltage-controlled amplifier, 48
  voltage-controlled oscillator, 99-100
impedance
  CA3080, 15
  op amps, 8
impedance matching
  buffers for, 31
  op amps for, 7
inverting amplifier
  ac-coupled, 36-40
  inverting Norton amplifier vs., 130
  LM3900 Norton amplifier, 139
  low-power, 40-42, 40
  Norton amplifier, 128-130, 145-148
  op amp in, 5
  polarity in, 31
  transconductance op amps in, 31
  variable gain, 43-44
inverting switch, fast, 58-60
inverting voltage comparator, 71-73
  gain in, 72-73
  resistor values, 72

L

LED flasher, dual, 203-206
linearizing diodes
  LM13600 dual OTA, 28, 29
  negative feedback through, 29

Index

LM13600 dual OTA, 24-29
  AGC amplifier using, 45
  amplitude modulator, 78-79, 78
  block diagram of, 27
  buffers in, 24, 25, 26
  Darlington transistors, 24, 25, 26
  gain, 29
  high-pass voltage-controlled filter, 119-121
  I-bias, 28, 29
  linearizing diodes in, 28, 29
  low-pass voltage-controlled filter, 117-119
  pin-out diagram and explanation, 24-25
  power supply for, 24
  ring modulator, 82-84
  voltage-controlled amplifier using, 51-53
LM3900 Norton amplifier, 133-143
  bandwidth, 136
  biasing current, 136
  circuit diagram for, 137, 138
  current mirrors (CM) in, 142-143
  current sink, improved version, 179-181
  current sinks, 177-181
  differential amplifier, 151-155
  dual LED flasher, 203-206
  fixed current source, 174-177
  flip-flop, 185-186
  four-input AND gate, 183-185
  frequency compensation, 136
  function generator, 202-203
  gain in, 139-140
  high-pass filter using, 157-161
  high-temperature alarm, 189-191
  input currents, 136
  internal structure, 136-143
  inverting amplifier circuit, 139, 141-148
  low-temperature alarm, 186-189
  narrow-width pulse generator, 196-198
  noninverting amplifier, 148-151
  output voltage, 136
  pin connections, 136, 137
  power supplies for, 135, 137
  Schmitt trigger, 170-171
  short-circuit protection, 136
  specifications for, 135-136
  square-wave generator, 191-194
  square-wave generator, alternative version, 194-195
  transistors in, 139-141
  under-voltage indicator, 173-174
  variable duty-cycle rectangle wave generator, 198-202
  variable voltage regulator, 165-167
  variable-reference voltage source, 167-168
  voltage comparator, 171-173
  voltage regulator, 163-167
  wide-bandwidth/high-gain amplifier, 155-157
load impedance, low-power inverting amplifier, 41, 42
load isolation, buffers for, 31
load resistance, current vs., 110
logic gates (see Page gates)
low-pass filters, 112, 114, 158
low-pass voltage-controlled filter, 111-119
low-power inverting amplifer, 40-42
  gain in, 40, 41
  I-bias in, 42
  load impedance in, 41, 42
  open-loop configuration, 42
  polarity inversion with, 41
low-power Schmitt trigger, 63-64
low-temperature alarm, Norton amplifier, 186-189

M

modulation, 55-89
  amplitude, 75
  amplitude, CA3080 OTA, 75-77
  amplitude, LM13600 OTA, 78-79
  amplitude, sideband generation, 79-82
  carrier and program signal in, 75
  carrier wave, 80, 81
  four-quadrant multiplier, 82, 84-87
  frequency in, 75, 80, 81
  fundamentals, 80, 81
  harmonics, 80, 81
  phase-amplitude modulator, 87-89
  program and carrier signals in, 55
  program wave, 80, 81
  pulse, 75

modulation (*cont.*)
  ring modulator, 82-84
  sine waves, 80, 81
  two-quadrant multiplier, 75
multipliers
  four-quadrant, 82, 84-87
  two-quadrant, 75
music synthesizers
  random music maker, 100-103
  sample-and-hold circuits, 70
  voltage-controlled amplifier for, 47, 49

N

narrow-width pulse generator, 196-198
negative feedback
  AGC amplifier, 46
  CA3080 OTA and, 40
  linearizing diodes and, 29
  Norton amplifier, 128
  op amps, 23
  transconductance op amps, 24
negative gain (*see* attenuation)
noise
  dead zones and, 65, 170
  hysteresis vs., 65, 170
  Schmitt trigger, 62, 63, 65, 169
  Schmitt trigger to clean up, 60-61
  voltage-controlled amplifier for, 47
noninverting amplifier
  Norton amplifier, 131-132, 148-151
  op amp in, 6
  polarity in, 32
  transconductance op amps in, 31, 32
noninverting voltage comparator, 73-75
  I-bias in, 74
  power consumption by, 73, 75
Norton amplifier, 121, 123-206
  amplifier projects, 145-161
  applications for, 127, 133
  basic principles of, 125
  bias current in, 129
  biasing, 127-128
  current differencing amplifier action of, 137
  current sink, 177-181
  current-differencing amplifier, 125

currents in, 129, 130
differential amplifier, 151-155
dual LED flasher, 203-206
feedback current in, 129, 130
fixed-current source, 174-177
flip-flop, 185-186
four-input AND gate, 183-185
function generator, 202-203
gain, 130, 132
high-pass filter, 157-161
high-temperature alarm, 189-191
input current in, 129, 130
internal circuitry of, 128, 129
inverting amplifier, 130, 145-148
inverting circuits, 128-130
LM3900, 133-143
low-temperature alarm, 186-189
narrow-width pulse generator, 196-198
negative feedback, 128
noninverting amplifier, 148-151
noninverting circuits, 131-132, 131
power supplies for, 126
precision current sinks, 133
precision current sources, 133
rectangle wave generator, 133
schematic symbol for, 125, 126
Schmitt trigger, 168-171
square-wave generator, 191-194
square-wave generator, alternative version, 194-195
under-voltage indicator, 173-174
unity gain, 132
variable duty-cycle rectangle wave generator, 198-202
variable voltage regulator, 165-167
variable-reference voltage source, 167-168
voltage and current projects, 163-181
voltage comparator, 133, 171-173
voltage-controlled oscillators using, 133
voltage regulator, 133, 163-167
wide-bandwidth/high-gain amplifier, 155-157
notch filter (*see* band-reject filters)

O

Ohm's law, 10, 11

Index   213

op amps
  applications for, 11
  buffer using, 7
  circuit design, resistor values, 8
  clipping, 23
  differential amplifier using, 7, 9
  frequency compensation in, 23
  gain in, 4, 6, 7, 8
  impedance in, 8
  impedance matcher, 7
  inverting amplifier using, 5
  inverting and noninverting inputs, 4, 5
  negative feedback, 23
  noninverting amplifier using, 6
  Norton amplifier vs., 125
  operation of, 4-8
  phase shifting in, 23
  RC network for phase shift control in, 23
  saturation, 23
  schematic symbol for, 5, 125, 126
  transconductance op amps vs., 4
  unity gain in, 6, 7
  voltage differencing amplifier action of, 137
open-loop configuration
  ac-coupled inverting amplifier, 39
  direct-coupled differential amplifier, 36
  low-power inverting amplifier, 42
oscillators, 91
  astable multivibrator, 191-194
  rectangle wave generator, 69-70
  voltage controlled, 11, 97-100
output chatter, Schmitt trigger, 62, 63, 65, 169

P

phase-locked loop, random music maker using, 100
phase modulation, phase-amplitude modulator, 87-89
phase shifting
  op amps, 23
  RC network to control, 23
phase-amplitude modulator, 87-89
polarity
  differential amplifier, 32
  inversion of, 6

  inversion of, buffers for, 31
  inverting amplifier, 31
  low-power inverting amplifier, 41
  noninverting amplifier, 32
positive feedback, 91
potentiometer
  random music maker, 103
  variable-gain inverting amplifier, 43-44
  variable-hysteresis Schmitt trigger, 66
power consumption
  low-power Schmitt trigger, 64
  noninverting voltage comparator, 73, 75
  transconductance op amps, 10, 12, 40
power supplies
  battery, under-voltage indicator for, 173-174
  LM13600 dual OTA, 24
  LM3900 Norton amplifier, 135, 137
  Norton amplifier, 126
  variable voltage regulator, 165-167
  variable-reference voltage source, 167-168
  voltage regulators for, 163
precision current sinks, Norton amplifier for, 133
precision current source, 110-111
  Norton amplifier for, 133
program signal, 55, 75, 80, 81
pulse generator, narrow-width, 196-198
pulse modulation, 75

R

random music maker, 100-103
  phase locked loop in, 100
  potentiometer in, 103
  sample-and-hold circuit, 102
  trigger frequencies for, 103
  voltage-controlled oscillator, 102
RC networks, phase shift control, op amps, 23
rectangle wave generator
  Norton amplifier for, 133
  sample-and-hold circuit, 69-70
  variable duty-cycle, 94-97, 198-202

rectangle-wave generator
  Norton amplifier, 191-194
  Norton amplifier, alternative version, 194-195
  pulse waves and, 196
reference voltages, 71
  variable, source for, Norton amplifier, 167-168
regulators, voltage (see voltage regulators)
remote control
  voltage-controlled amplifier for, 47
  voltage-controlled resistance and, 108-110
resistance, voltage controlled, 108-110
resistors
  amplitude modulator, CA3080, 77
  analog switch, 57
  fast inverting switch, 59
  inverting voltage comparator, 72
  op amp circuit design, 8
  Schmitt trigger, 64
  variable-hysteresis Schmitt trigger, 67
  voltage-controlled amplifier, 48
ring modulator, 82-84
ripple, 163

S

sample-and-hold circuit, 11, 67-70
  A/D conversion with, 70
  buffer amp for, 68-69
  capacitor selection for, 69
  clock or control pulse generator, 69
  data recording with, 70
  I-bias in, 69
  music synthesis with, 70
  op amp selection for, 69
  random music maker, 102
  signal transmission with, 70
saturation, op amps, 23
Schmitt trigger, 60-67
  analog to pulse wave conversion with, 61, 62
  comparator use of, 61
  dead zones and, 65, 170
  hysteresis in, 65, 170
  I-bias in, 62
  low-power, 63-64

noise from, 62, 63, 65, 169
Norton amplifier, 168-171
output chatter from, 62, 63, 65, 169
power consumption, 64
resistor values, 64
variable-hysteresis, 64-67
sidebands, 79-82
signal generation projects, 91-103
  function generator, 202-203
  narrow-width pulse generator, 196-198
  random music maker, 100-103
  square-wave generator, 91-94
  square-wave generator, Norton amplifier, 191-195
  variable duty-cycle rectangle wave generator, 94, 198
  voltage-controlled oscillator, 97-100
signal transmission, sample-and-hold circuit for, 70
sine waves, 75, 80, 81
slew rate
  CA3080 OTA, 16, 39, 59
  square-wave generator, 92
speed-up diodes, CA3080 OTA, 20
SPST switch, analog switch as, 57
square wave, 94
  generator for, 91-94
  Norton amplifier, 191-194
  Norton amplifier, alternative version, 194-195
  VCO to generate, 98, 99
stereo audio, voltage-controlled amplifier for, 53
switching, 11, 55-89
  analog switch, 55-58
  fast inverting switch, 58-60
  inverting voltage comparator, 71-73
  low-power Schmitt trigger, 63-64
  noninverting voltage comparator, 73-75
  sample-and-hold circuits, 67-70
  Schmitt trigger, 60-67
  variable-hysteresis Schmitt trigger, 64-67

T

transconductance (see also gain), 3-4
transconductance op amps
  ac-coupled inverting amplifier, 36-40

Index

AGC amplifier, 44-46
amplifier using, 31-53
amplitude modulator, CA3080 OTA, 75-77
amplitude modulator, LM13600, 78-79
analog switch, 55-58
applications for, 10-12
CA3080, 13-24
current drain, 40
definition, 3
differential amplifier using, 9, 31, 32
differential amplifier, direct-coupled, 32
fast inverting switch, 58-60
frequency compensation and, 24
gain in, 9, 10
harmonics remover, 105-108
high-pass voltage-controlled filter, 119-121
I-bias in, 4, 9, 10, 12
inverting amplifier using, 31
inverting voltage comparator, 71-73
LM13600 dual, 24-29
low-pass voltage-controlled filter, 111-119
low-power inverting amplifier, 40-42
low-power Schmitt trigger, 63-64
modulation with, 55-89
negative feedback and, 24
noninverting amplifier using, 31, 32
noninverting voltage comparator, 73-75
Norton amplifier (see Norton amplifier)
Norton amplifier vs., 125
op amps vs., 4
open-loop configuration, 36, 39, 42
operation of, 9-10
outputs of, 9
phase-amplitude modulator, 87-89
power consumption, 10, 12, 40
precision current source, 110-111
random music maker, 100-103
ring modulator, 82-84
sample-and-hold circuits, 11, 67-70
schematic symbol for, 9
Schmitt trigger, 60-67
signal-generation projects, 91-103
square-wave generator, 91-94
switching applications, 11, 55-89
triangle-to-sine wave conversion, 11
variable duty-cycle rectangle wave generator, 94-97
variable-gain inverting amplifier, 43-44
variable-hysteresis Schmitt trigger, 64-67
voltage applications, load resistors in circuit of, 11
voltage-controlled amplifier, 11, 47-53
voltage-controlled filters, 11
voltage-controlled oscillator, 11, 97-100
voltage-controlled resistance, 108-110
transistors, CA3080 OTA, 20
triangle wave, VCO to generate, 98, 99
triangle-to-sine wave conversion, 11
trigger frequencies, random music maker, 103
truth tables, 183
two-quadrant multipliers, 75

U

under-voltage indicator, 173-174
unity gain, 6, 7
  buffers and, 31
  Norton amplifier, 132

V

variable duty-cycle rectangle wave generator, 94-97, 198-202
variable-gain inverting amplifier, 43-44
variable-hysteresis Schmitt trigger, 64-67
  calibration of, 66
  I-bias, 65-66
  potentiometer in, 66
  resistor values, 67
variable-reference voltage source, Norton amplifier for, 167-168
voltage and current projects, Norton amplifier, 163-181

voltage comparator
  inverting, 71-73
  noninverting, 73-75
  Norton amplifier, 133, 171-173
voltage-controlled amplifier, 11, 47-53
  amplitude modulator vs., 76
  analog switch vs., 55-58
  applications for, 47
  automation applications for, 47
  CA3080 OTA for, 47
  calibration of, 49
  I-bias, 48
  improved version, 50-51
  LM13600 OTA for, 51-53
  music synthesis with, 47, 49
  noise reduction with, 47
  nonlinearity in, 50
  op amp selection for, 49, 51
  potentiometer for dc offset adjustment, 49
  remote control with, 47
  resistor values, 48
  simple configuration, 47-50
  stereo audio applications, 53
  voltage divider in, 48
voltage-controlled filter, 11, 105
  high-pass, 119-121
  low-pass, 111-119
voltage-controlled oscillator, 11, 97-100
  I-bias, 99-100
  Norton amplifier for, 133
  random music maker, 102
  square and triangle wave generation, 98, 99

voltage-controlled resistance, 108-110
voltage differencing amplifier, 137
voltage divider
  ac-coupled inverting amplifier, 38
  direct-coupled differential amplifier, 34
  voltage-controlled amplifier, 48
voltage regulators
  Norton amplifier for, 133, 163-167
  variable, 165-167
voltage sources, reference, variable, Norton amplifier, 167-168

W

waveforms
  analog to pulse form conversion, Schmitt trigger for, 61, 62
  duty cycles, 94, 95
  function generator for, 202-203
  fundamentals, 80, 81, 95, 115, 199, 200
  harmonics, 80, 81, 95, 115, 199, 200
  harmonics remover circuit, 105-108
  modulation and, 75
  pulse waves, 196-198
  sine waves, 75, 80, 81
  square wave, 94, 98, 99
  triangle wave, 98, 99
  triangle-to-sine conversion, 11
wide-bandwidth/high-gain amplifier, 155-157

Z

zener diodes, 163

Other Bestsellers of Related Interest

TROUBLESHOOTING AND REPAIRING VCRs
—Gordon McComb

It's estimated that 50% of all American households today have at least one VCR. *Newsweek* magazine reports that most service operations charge a minimum of $40 just to look at a machine, and in some areas there's a minimum repair charge of $95 *plus the cost of any parts*. Now this time- and money-saving sourcebook gives you complete schematics and step-by-step details on general up-keep and repair of home VCRs—from the simple cleaning and lubricating of parts, to troubleshooting power and circuitry problems. 336 pages, 300 illustrations. Book No. 2960, $27.95 hardcover, $17.95 paperback

ALARMS: 55 Electronic Projects and Circuits
—Charles D. Rakes

Make your home or business a safer place to live and work—for a price you can afford. Almost anything can be monitored by an electronic alarm circuit—from detecting overheating equipment to low fluid levels, from smoke in a room to an intruder at the window. This book is designed to show you the great variety of alarms that are available. There are step-by-step instructions, work-in-progress diagrams, and troubleshooting tips and advice for building each project. 160 pages, 150 illustrations. Book No. 2996, $19.95 hardcover, $12.95 paperback

49 BATTERY-POWERED TWO-IC PROJECTS
—Delton T. Horn

Using detailed diagrams, parts lists, and step-by-step instructions, Horn shows you how to build useful two-IC devices including: delayed trigger timer, power amplifier, light range detector, digital filter and other inexpensive projects. These projects are perfect for developing the skills and the confidence you need to tackle more involved projects. Even if you're already building robots and computers, you can certainly enjoy the relaxation of whipping together a quick and simple device. 126 pages, 125 illustrations. Book No. 3165, $17.95 hardcover, $10.95 paperback

MAINTAINING AND REPAIRING VCRs
2nd Edition—Robert L. Goodman

" . . . of immense use . . . all the necessary background for learning the art of troubleshooting popular brands" said *Electronics for You* about the first edition of this indispensable VCR handbook. Revised and enlarged, this illustrated guide provides complete, professional guidance on troubleshooting and repairing VCRs from all the major manufacturers, including VHS and Betamax systems and color video camcorders. Includes tip on use of test equipment and servicing techniques plus case history problems and solutions. 352 pages, 427 illustrations. Book No. 3103, $27.95 hardcover, $17.95 paperback

101 OPTOELECTRONICS PROJECTS
—Delton T. Horn

Discover the broad range of practical applications for optoelectronic devices! Here's a storehouse of practical optoelectronics projects just waiting to be put to use! Horn features 101 new projects including: power circuits, control circuits, sound circuits, flasher circuits, display circuits, game circuits, and many other fascinating projects! This book offers you an opoportunity to make a hands-on investigation of the practical potential of optoelectronic devices. 240 pages, 273 illustrations. Book No. 3205, $24.95 hardcover, $16.95 paperback

44 POWER SUPPLIES FOR YOUR ELECTRONIC PROJECTS
—Robert J. Traister and Jonathan L. Mayo

Here's a sourcebook that will make an invaluable addition to your electronics bookshelf whether you're a beginning hobbyist looking for a practical introduction to power supply technology, with specific applications . . . or a technician in need of a quick reference to power supply circuitry. You'll find guidance in building 44 supply circuits as well as how to use breadboards, boards, or even printed circuits of your own design. 220 pages, 208 illustrations. Book No. 2922, $24.95 hardcover, $15.95 paperback

THE LASER COOKBOOK: 88 Practical Projects
—Gordon McComb

The laser is one of the most important inventions to come along this half of the 20th Century. This book provides 88 laser-based projects that are geared toward the garage-shop tinkerer on a limited budget. The projects vary from experimenting with laser optics and constructing a laser optical bench to using lasers for light shows, gunnery practice, even beginning and advanced holography. 400 pages, 356 illustrations. Book No. 3090, $25.95 hardcover, $18.95 paperback

BASIC ELECTRONICS THEORY—3rd Edition
—Delton T. Horn

"All the information needed for a basic understanding of almost any electronic device or circuit . . ." was how *Radio-Electronics* magazine described the previous edition of this now-classic sourcebook. This completely updated and expanded 3rd edition provides a resource tool that belongs in a prominent place on every electronics bookshelf. Packed with illustrations, schematics, projects, and experiments, it's a book you won't want to miss! 544 pages, 650 illustrations. Book No. 3195, $28.95 hardcover, $21.95 paperback

Prices Subject to Change Without Notice.

500 ELECTRONIC IC CIRCUITS WITH PRACTICAL APPLICATIONS—James A. Whitson

More than just an electronics book that provides circuit schematics or step-by-step projects, this complete sourcebook provides both practical electronics circuits AND the additional information you need about specific components. You will be able to use this guide to improve your IC circuit-building skills as well as become more familiar with some of the popular ICs. 336 pages, 600 illustrations. Book No. 2920, $29.95 hardcover, $19.95 paperback

CUSTOMIZE YOUR PHONE: 15 Electronic Projects—Steve Sokolowski

Practical, fun, phone enhancement projects that anyone can build. A melody ringer, an automatic recorder, and a telephone lock—these are just a few of the improvements you can add to make your everyday telephone more interesting and more useful. Steve Sokolowski explains the basics of constructing an electronic project as well as the fundamentals that make your telephone work. While the projects are all rather simple and inexpensive (built for about $10 to $30 each) they are also very useful. 176 pages, 125 illustrations. Book No. 3054, $19.95 hardcover, $12.95 paperback

---

Look for These and Other TAB Books at Your Local Bookstore

To Order Call Toll Free 1-800-822-8158
(in PA, AK, and Canada call 717-794-2191)

or write to TAB BOOKS, Blue Ridge Summit, PA 17294-0840.

Title	Product No.	Quantity	Price

☐ Check or money order made payable to TAB BOOKS

Charge my ☐ VISA ☐ MasterCard ☐ American Express

Acct. No. _____ Exp. _____

Signature: _____

Name: _____

Address: _____

City: _____

State: _____ Zip: _____

Subtotal $ _____

Postage and Handling
($3.00 in U.S., $5.00 outside U.S.) $ _____

Add applicable state
and local sales tax $ _____

TOTAL $ _____

TAB BOOKS catalog free with purchase; otherwise send $1.00 in check or money order and receive $1.00 credit on your next purchase.

Orders outside U.S. must pay with international money order in U.S. dollars.

TAB Guarantee: If for any reason you are not satisfied with the book(s) you order, simply return it (them) within 15 days and receive a full refund.

BC